3.

GEARS AND THEIR
VIBRATION

MECHANICAL ENGINEERING

A Series of Textbooks and Reference Books

EDITORS

L. L. FAULKNER

Department of Mechanical Engineering
The Ohio State University
Columbus, Ohio

S. B. MENKES

Department of Mechanical Engineering
The City College of the
City University of New York
New York, New York

OTHER VOLUMES IN PREPARATION

GEARS AND THEIR VIBRATION

A Basic Approach to Understanding Gear Noise

J. DEREK SMITH

University Engineering Department
Cambridge University
Cambridge, England

MARCEL DEKKER
THE MACMILLAN PRESS LTD.

New York and Basel
London and Basingstoke

Library of Congress Cataloging in Publication Data

Smith, James D. (James Derek),
   Gears and their vibration.

   (Mechanical engineering ; 17)
   Includes index.
   1. Gearing--Noise. 2. Gearing--Vibration. I. Title.
II. Series.
TJ184.S58 1983   621.8'33   82-22126
ISBN 0-8247-1759-7
ISBN 0-333-35045-6 (British ISBN)

Distributed in the UK, Europe, and Africa by The Macmillan Press
Ltd., London and Basingstoke.

MARCEL DEKKER, INC.
270 Madison Avenue, New York, New York 10016

Current printing (last digit):
10   9   8   7   6   5   4   3   2   1

PRINTED IN THE UNITED STATES OF AMERICA

PREFACE

The objective of this book is to help engineers who encounter gear noise problems to understand the mechanisms of noise generation and to tackle gear vibration problems in an ordered manner.

Direct production of noise by gears is rare; a point emphasized in the book is that vibration travels through the gearcase and surrounding structure and is eventually radiated as noise.

I have tried to balance the gear aspects which are needed for the vibration engineer and the vibration aspects which are needed for the gear engineer. In the process I hope that this book will be suitable for those who have little experience of either gears or vibration; both subjects involve some very simple concepts which are usually wrapped up in incomprehensible jargon and irrelevant mathematics.

To those who already have experience of gears or vibration some chapters are redundant. My attempts to compress aspects such as manufacture, measurement, or analysis into a short chapter will amuse authors who have not managed to fit such large topics into complete books. Some chapters approach familiar subjects from the point of view of noise rather than strength and so may shed a different light on well known facts.

In general, details of particular mathematical approaches such as computational methods for frequency analysis have been omitted because they are not relevant for those whose objective is the elimination of practical troubles and gear subtleties such as correction and undercutting have been ignored. Inevitably the compression of so much material into one book will not suit all tastes.

This book owes much to Mr. D. B. Welbourn of the Wolfson Industrial Liaison Unit at Cambridge. He originally introduced me to gearing and has consistently helped and supported me and broadened my experience of gears and their problems and behavior. I have also learned much from many friends in industry and academic life who have shared their knowledge and expertise. I am also much indebted to Dr. H. Kohler and Dr. A. Thompson whose comments and suggestions were extremely helpful.

J. Derek Smith

CONTENTS

APPENDIX

INDEX

GEARS AND THEIR VIBRATION

Chapter 1

INTRODUCTION

I. DEVELOPMENT OF GEARING

The earliest drives used cylindrical rods inserted radially in one wheel meshing with similar rods mounted axially in another wheel as sketched in Figure 1.1. This type of drive performed satisfactorily, though inefficiently, at low speeds and low loads but trouble was encountered as soon as load or speed was raised.

The most obvious limitation was that of load. The contact is theoretically a point contact giving very high local stresses which the materials could not withstand and ensuring that any lubrication was ineffective so high wear occurred. Less obvious, but eventually more important was the fact that the speed ratio was not constant so that if one gear ran at a constant speed, the other gear had to accelerate and decelerate each "tooth." The loads associated with the accelerating forces were proportional to inertia times speed squared and quickly dominated the steady drive loads to cause failure and vibration.

The first half of the 19th century saw the first effective gear cutting machine (1) which could generate the accuracies required and mathematicians such as Willis (2) pointed out the advantages of involutes in preference to cycloidal shapes.

The subsequent history of gear drives has been concerned mainly with keeping the contact stresses below material limits and with improving the smoothness of drive, i.e., keeping the velocity ratio as constant as possible to avoid dynamic effects which will give stress increases, vibration, and noise. It is only with the last few years that there has been enough understanding of the detailed operation of gears to give the extremely smooth drives which are now necessary. As a rough guide the velocity ratio of a drive should not vary by more than 0.01% for quiet operation.

Opitz (3) published interesting curves showing the variation of noise from many gearboxes as a function of power and gearbox accuracies. Output sound power varied between 10^{-8} of the transmitted power and 10^{-6} of the transmitted power. This, at 4 m distance from a 1000 kW geardrive gives a variation from 80 dB to over 100 dB in sound pressure level.

1

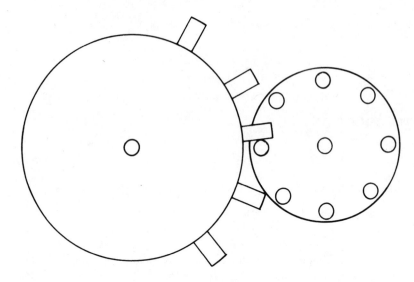

Figure 1.1.   Sketch of early drive system. Rods inserted radially in the wheel mesh with rods fitted axially between two discs.

II.   VELOCITY RATIO REQUIREMENT

Constant velocity ratio of a gear drive is so important that the first essential is to find the "geometric" rule for constant ratio. The two shapes shown in Figure 1.2 are in contact at X and are rotating about centers O and Q. Whatever the shapes, there must be a common tangent at X and a common normal XN. Provided the "shapes" remain in contact they must have the same velocity in the direction of the normal XN so that $OA \times w_1 = QB \times w_2$.

   Putting in P, the point where the instantaneous common normal XN intersects the line of centres and defining ϕ as the angle between the normal and a line perpendicular to OQ gives $OP \cos \phi \times w_1 = QP \cos \phi \times w_2$, i.e.,

$$\frac{w_1}{w_2} = \frac{QP}{OP}$$

So the only requirement for a constant velocity ratio is that the ratio of QP:OP remains constant which means that the common normal at the point of contact must pass through the fixed point P which is usually called the pitch point. There are an infinite number of pairs of shapes which will meet this requirement and there have been a very large number of ingenious ideas which are unsatisfactory.

   We may also deduce from Figure 1.2 the relative sliding velocity at the contact point as $PXw_1 + PXw_2 = PX(w_1 + w_2)$; the theoretical power loss is then the friction force multiplied by this sliding velocity.

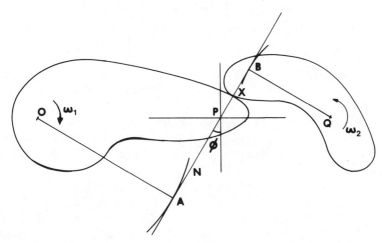

Figure 1.2. Geometry of meshing. The common normal XN at the point of contact X cuts the line joining centers O and Q at the point P. QB and OA are perpendicular to XN.

III. CHOICE OF INVOLUTE

The involute came into general use as the preferred gear profile primarily because the meshing action of the involute is not disturbed by small errors in the center distance of the gears. For involute gears the common normal PN is best thought of as a string which is unwrapping from a disc of radius OA and reeling onto a disc of radius QB. The point X is a fixed point on the string and so describes an involute relative to QB and a (different) involute relative to OA as it in reality travels from A to B. The force is always along the string.

Movement of O and Q apart will alter the angle at which the string AB lies but will not disturb the smoothness of mesh. The angle ϕ (the pressure angle) will alter from say $20°$ to $21°$ but the geometry is correct. It is common usage in gearing to alter center distances and then gears which are nominally $20°$ pressure angle gears are not actually working at $20°$ pressure angle.

It should perhaps be noted that the unwrapping string concept on which all involute gearing geometry is based requires a rather curious string which is perfectly flexible when wrapped round the discs OA and QB but which acts rigidly in compression between A and B.

Sensitivity to center distance is a major disadvantage of all the gear forms based on cycloids and of the more modern Wildhaber-Novikov gears though manufacturing improvements have mitigated this disadvantage. There is, however, another disadvantage of non-involute gears in that although other gears can be designed so that the common normal and, hence, the force always passes through the point P, only involute gears allow the normal and the force to keep acting in the same direction as well as at the same place. In general, noise or vibration will be produced in a structure if either the size of a force varies with time, or the point of application of a force varies or the direction of the force varies.

Cycloidal gears allow the direction of the force to vary and, hence, must generate more vibration than involute gears. Wildhaber-Novikov gears arrange that the contact only occurs at P and pressure is in one direction only (usually perpendicular to the line joining centers) but suffer from the corresponding disadvantage that the point of application of the force must move axially to maintain drive and this leads to axial variations of force position which give noise and vibration.

The involute has remained as the favored gear profile till now primarily due to reasons of ease of manufacture and tolerance of errors but its use may now be justified purely on vibration grounds as it is the only profile which can give constant force, force direction, and force position for applications where lack of vibration is essential.

IV. COMPARISONS WITH OTHER DRIVES

There are many competitors to gear drives so it is worthwhile checking whether or not a gear drive is the most suitable method of transmitting power. The main competitors are belt drives which can be flat, V, or toothed, chain drives, and roller or inverted-tooth and friction drives which are making a comeback for variable speed applications. Particular applications may be best suited by hydraulic or electric power transfer which have the asset of allowing some energy storage and can give variable speed. Gear drives

    i. are extremely compact and, hence, useful when space is limited but belt or hydraulic are better for long transfers.

    ii. need lubrication. Can be an advantage or disadvantage.

    iii. are extremely rigid. Can be an advantage or disadvantage.

    iv. give synchronous drive. An exact ratio is necessary for many applications but belt creep or hydraulic slip can help with load sharing.

    v. give direction reversal (usually).

    vi. do not tolerate misalignments if loadings are high.

    vii. are very accurate, typically within $20\mu m$ at pitch radius where a toothed belt drive may have $200\mu m$ errors.

    viii. are much cheaper than hydraulic or electric drive for high torques.

    ix. are quieter than other drives for high torque applications provided accurate gears are used but are likely to be noisier at very low torques when loss of contact occurs.

    x. can run at very high speeds, unlike belt drives which do not like running faster than 200 m/min.

The ultimate control on a decision is, of course, cost; gears usually are at an advantage for high torque applications such as ships propeller final drives but may be uncompetitive for very low torques.

REFERENCES

1. Dudley D. W., The Evolution of the Gear Art, American Gear Manufacturers Association, 1969.
2. Willis R., On the Teeth of Wheels, Trans. Institute of Civil Engineers, Vol. II, 1838.
3. Opitz H., Noise of Gears, Phil. Trans. of Royal Society, Vol. 263, 1969, pp. 369–380.

Chapter 2

SPUR GEARS

I. DEFINITIONS

Spur gears are gears in which the teeth lie parallel to the axis of rotation. We are dealing solely with the theory of parallel axis gears, and so any section perpendicular to the axes of a pair of spur gears will be exactly the same as any other section.

The simple theory in chapter 1 allows us to consider the motion as the same as that of a string unwrapping from one disc and reeling onto another with a particular point on the string describing an involute relative to each disc. This is shown in Figure 2.1; OA and QB are the radii of the discs which are called base circles and the string is straight between points A and B whose positions are given by the points where the line of pressure, i.e., the string, is tangential to the two base circles. Any point X on the string describes an involute relative to each base circle and so defines the shape of the tooth flank.

The point P where the line of pressure alternatively called the line of action, crosses the line joining the centers OQ is the pitch point and we define pitch circle radii as OP and QP. We also define the pressure angle ϕ as the angle between the line of pressure and the pitch line which is the common tangent to the two pitch circles, perpendicular to the line OQ. It is important to realize that the base circle radius r_b is an absolute property of a gear and is fixed once a gear is manufactured whereas the pressure angle ϕ and the pitch circle radius r_p are nominal only and will alter if the gear center distance is altered. They are linked by the relationship that $r_p \cos\phi = p_b$ which is fixed for any particular wheel. It is however more convenient to specify r_p and ϕ as if they were absolute properties of the gear so they are usually quoted to describe a gear; the pitch circle is then called a reference circle and corresponds to the nominal pressure angle (usually $20°$).

The point X on the "string" will generate an involute which will define one gear tooth flank on a gear and must be followed by successive points on the string to generate the remainder of the teeth on the gear wheel.

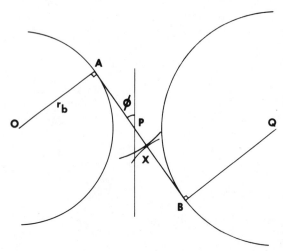

Figure 2.1.   Involute geometry. AB is the unwrapping string which represents the common normal at the point of contact X.

The distance between successive teeth may be defined as the distance round the pitch circle which is $2\pi r$ divided by N the number of teeth; this is called circumferential pitch p. Alternatively it is defined as the distance between successive teeth measured along the pressure line and is called base pitch p_b and is equal to the distance round the base circle between successive teeth; this is $2\pi r_b$ divided by N the number of teeth.

II. STANDARDIZATION

Some gears require an exact pitch circle circumference in equipment such as printing machinery and so it is logical to specify pitch as an exact distance (e.g., 0.5 inches) for such gears. This requirement is unusual however and so most gears are specified in terms of a *module* m which is the pitch circle diameter (in mm) divided by the number of teeth. (The old system similarly links diameter and number of teeth but the other way up, thus giving the number of teeth per inch of diameter of pitch circle and is called diametral pitch).

The center distance of a pair of gears is then rapidly determined; if a 52 tooth 2 mm module gear meshes with a 22 tooth gear, the pitch circle diameters are 104 and 44 mm so the center distance is 74 mm. The pitch is simply π m, i.e., 2π mm in this case.

The nominal pressure angle ϕ requires standardization and although other angles such as 14½ degrees are occasionally used, industry has fixed on $20°$ pressure angle.

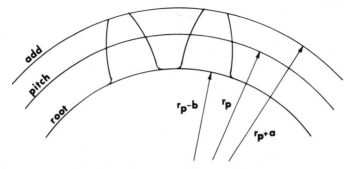

Figure 2.2. Sketch showing gear radii. a is the addendum and b is the dedendum.

There is also a requirement to decide how far a gear tooth will project above the pitch circle radius (the addendum) and to what depth the roots of the teeth should be cut below the pitch circle radius (the dedendum). Figure 2.2 shows the addendum and dedendum circles; it is usual to have the addendum equal to the module and the dedendum some 15% greater than the module.

When teeth are "corrected" the position of the tooth relative to the pitch circle is moved and it is possible to have teeth which are all addendum meshing with teeth which are all dedendum. This is sometimes an advantage for clearance or to reduce interference at teeth roots but it has the disadvantage that sliding velocities are increased as all the contact takes place one side of the pitch point.

Only one flank of each tooth has been considered so far but it is normal to make teeth symmetrical for ease of manufacture or for conditions where the other flank is used, e.g., for overrun in a vehicle or in an idler gear which uses both flanks. Theoretically, for drives which use only one flank it is possible to obtain a stronger tooth by making it asymmetric but this is virtually never done.

Clearance should be left between the non-working flanks of a pair of mating gears, otherwise forces and wear rates are high; this clearance is called backlash and is usually not less than 100 μm (0.004 inches) though precision or servo drives may require lower values. Tooth thicknesses are usually roughly equal, that is the tooth thicknesses at the pitch (reference) circle radius, although to match root bending strengths of teeth, gears with small numbers of teeth should have thicker teeth at the pitch circle than gears with large number of teeth.

III. CONTACT RATIO

When a pair of spur gears mesh we know that the point of contact between the mating involutes travels along the line of pressure (action). It is obviously important to ensure that one pair of teeth maintain contact until the next pair of teeth are ready to take load. Less obvious from simple theory is that the greater

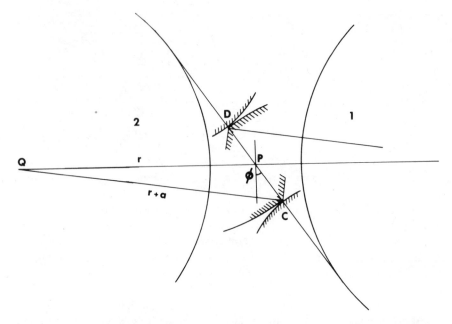

Figure 2.3.   Length of path of contact. CD is limited by loss of contact at the tip of 2 at C and of 1 at D. PC is calculated from ϕ and the tip and pitch circle radii of gear 2.

the "handover" time from one pair of teeth to the next the lower will be the noise and vibration level from the gears.

The term "contact ratio" is used to specify the average number of pairs of teeth in contact in a spur mesh and this ratio is of great importance in the design of both spur and helical gears. In Figure 2.3 a pair of teeth are shown when they have just made contact at C and the same pair of teeth are shown later when they are just about to lose contact at D. The distance CD is called the path of contact or length of action and if this distance CD is divided by the distance between successive working tooth flanks along the line of pressure (action), the ratio is the contact ratio. In most spur gears there are two pairs of teeth in contact for about half the time and one pair in contact for the remainder of the time.

The distance between successive flanks is already known because it is simply the base pitch p_b and is $\pi m \cos \phi$ since the base circle radius is $\cos \phi$ times the pitch circle radius. The requirement is to determine the length CD; the position of C is given by the limitation that the line of pressure (action) reaches the addendum of gear 2 while D is given as the point where the pressure line reaches the addendum of 1. In all normal gears contact ceases at the tip of the working flank and not at the root. This means that the distance PC is governed by the addendum of 2 and distance PD by the addendum of 1.

Figure 2.3 shows the simple triangle which connects the gear radii and the length PC. The standard triangle formula gives

$$(r + a)^2 = r^2 + (PC)^2 - 2, r \, PC \cos CPQ$$

$$\text{i.e., } r^2 + 2ar + a^2 = r^2 + (PC)^2 + 2 \, r \, PC \sin \phi$$

$$\text{and } PC = - r \sin \phi \pm [(r \sin\phi)^2 + (a^2 + 2ar)]^{\frac{1}{2}}$$

Only the positive sign is relevant since the negative sign gives the other intersection of pressure line and circle. PD is found by using the radii and addendum of gear 1 and the total path of contact or length of action is then known.

The contact ratio increases with the numbers of teeth in a gear pair but reaches a limiting value. For very large gears it appears visually that the contact ratio is high but the approach and recess lengths of contact (action) are m cosec ϕ since the line of action is inclined at angle ϕ and the addendum is m. The base pitch p_b is $\pi m \cos \phi$ and so the contact ratio is 2 cosec $\phi/\pi \cos \phi$ which for $20°$ pressure angle is 1.98. A contact ratio of over 2 can only be achieved by using a larger addendum than standard or by decreasing the pressure angle; these both give more slender teeth which are less strong in bending but are quieter.

Typical figures for a 24/45 tooth mesh with standard teeth are

$$PC = [(12 \, m \sin 20°)^2 + (1 + 2.1.12) \, m^2]^{\frac{1}{2}} -12 \, m \sin 20°$$

$$\text{and } PD = [(22.5 \, m \sin 20°)^2 + (1 + 2.1.22.5) \, m^2]^{\frac{1}{2}} - 22.5 \, m \sin 20°$$

giving CD = 2.36 + 2.56 m and, hence, a contact ratio of $4.92/\pi \cos 20°$, i.e., 1.667.

IV. THEORETICAL CONTACT STRESSES

Contact stresses are usually the limiting factor in gear design so estimation of contact stress levels is required. Figure 2.4 shows the "unwrapping string" with contact occurring over the path CD. At a point such as X the radii of curvature of the gear teeth are XE and XF so the mean Hertzian contact stress is given by

$$\bar{q} = \frac{1}{4} \left(\frac{P'\pi}{R} \cdot \frac{E}{2(1-\nu^2)} \right)^{\frac{1}{2}}$$

where E is the Young Modulus, ν is Poissons ratio and P' the load per unit width. The maximum compression stress is $\bar{q} \, 4/\pi$ and the maximum shear stress is about $0.38 \, \bar{q}$ and occurs below the surface (usually about 0.5 mm). R is the effective radius of curvature at the contact and is given by $1/R = 1/XE + 1/XF$. The distance EF is fixed so that XE + XF fixed and R is a maximum when the contact point is midway between E and F and a minimum when the contact distance CE

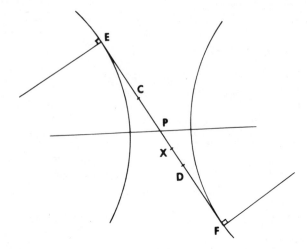

Figure 2.4. Radii of curvature of involutes. The centers of curvature are at E and F; the contact point X moves between C and D.

or DF is smallest. If P' is fixed, the stress varies continuously with variations in R; however P' varies due to the effects of contact ratio as 2 teeth come into mesh and the value of P' correspondingly halves at both ends of the path of contact. The realities of the variations in contact stresses are sufficiently complex when account has been taken of tooth deflection that it is more realistic, though rather arbitrary, to estimate contact pressures solely at the pitch point P and use these for comparison of different designs; at the pitch point there is only one pair of teeth in contact.

Taking the 24/45 drive in section III with a module of 3 mm (8DP) and a loading of 175 N/mm (1000 lbf/inch) gives PC = 12 × 3 × sin 20° and PD = 22.5 × 3 × sin 20° so for contact at the pitch point the effective R is 8.03 mm.

The mean contact pressure is $\bar{q} = \frac{1}{4} \left(\frac{175 \times 10^3 \times \pi}{8.03 \times 10^{-3}} \times \frac{210 \times 10^9}{2(1 - 0.3^2)} \right)^{1/2}$ which is

7.03×10^8 N/m² and gives a maximum stress of 8.95×10^8 N/m² corresponding to a strain of 0.00426.

V. ROOT STRESSES

Estimation of root stresses in a gear tooth is so complex due to the peculiar shape that theoretical attempts have not been successful; photoelastic methods and computed finite element methods are more suitable and have given agreed results which are now incorporated into specifications.

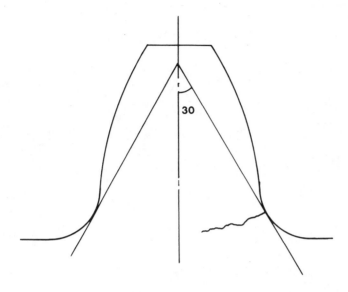

Figure 2.5. Tooth root stresses. The maximum stress is taken to occur where the 30° line touches the root.

The maximum stresses are tensile and occur below the working section of the flank in the root as shown in Figure 2.5; they give crack propagation from any stress raisers such as machining marks or surface defects and eventually break off the complete tooth. The level of stress is not greatly influenced by the position of the acting force on the flank and in practice the applied force decreases towards the tooth tip so that the most critical position is about one third of the way down the flank; the maximum stress position is assumed to occur where a tangent at 30° to a radial line meets the tooth. The level of stress has been measured and estimated for a particular case and correction is then applied for variations in tooth shape due to numbers of teeth, correction, change in root shape etc. A particular condition for a standard 20° pressure angle tooth with a very large number of teeth gave a value of 4.932 for the ratio of the tensile stress at the 30° tangent point to the force per unit facewidth divided by the module, i.e.,

$$\sigma_{30} = 4.932 \cdot \frac{P'}{m} \qquad \text{(The constant is called } Y_{FS}\text{)}$$

The lowest stress is encountered with about 60 teeth with a factor of 4.46 and by 18 teeth the factor has risen to 4.93 again. Further corrections must subsequently be made to allow for local stress-raising defects.

VI. SIZE EFFECTS

The estimates in sections IV and V give an idea of the levels of stresses but do not take into account local variations in loading cycles, dynamics, machining defects, accuracies, materials, etc. The more modern specifications such as DIN 3990 attempt to take into account such factors but can only do so when they are known.

A particular firm will have experience of a range of applications and will require prediction of performance of new designs without going back to fundamentals since they are often not known. The need is for a set of "scaling" rules which will allow extrapolation from existing knowledge. This section deals with "scaling" effects.

Root bending stresses are proportion to the load per unit facewidth P' divided by the module m for a given shape of tooth and are very little affected by number of teeth. In contrast the Hertzian contact stresses are proportional to the square root of P' divided by R the effective radius of curvature and are unaffected by tooth size i.e., module. Since material permissible stress is fixed we can say for:

Bending	Load/unit width	$m \times \sigma$	only
Contact	Load/unit width	$R \times \sigma^2$	only

The torque transmitted by a set of gears is the important factor so with facewidth b and number of teeth N, the relations become

Bending	Torque $\alpha\, b \times (Nm) \times m \times \sigma$
Contact	Torque $\alpha\, b \times (Nm)^2 \times \sigma^2$

Nm is proportional to diameter.

As might be expected, if contact stress limits, torque is proportional to $b \times d^2$, i.e., to volume and, hence, to weight but if bending (root) stress limits, torque is only proportional to volume if the number of teeth remains constant. If the size of tooth is increased the torque increases for a given volume. In a "balanced" design bending and contact stresses give the same permissible torque; there is however no inherent advantage in this so it is usual to have large teeth so that bending stresses are relatively low. A consequence of this is that it is often possible to increase tooth numbers in a gear, without increasing size, to raise vibration frequencies without altering permissible surface loadings since the diameters and hence the radii of curvature have not changed.

As a simple example of the effects of changes, consider a gearbox which is known to transmit 100 kW at 3000 rpm input with 3 mm module gears. What power should a similar design and manufacture gearbox transmit at 3600 rpm with:

   i. The same size of gearbox but with 4 mm module teeth?

  ii. The same number of teeth but 4 mm module and the same facewidth?

 iii. A geometrically similar gearbox scaled by 4/3?

 iv. The same gearbox with permissible material stresses increased by 20%?

The effect of the speed change, at constant torque, is to increase power to 120 kW; then for each case, *if contact stresses control* performance, the limits become:

   i. 120 kW. No change since radii unaltered.

  ii. $120 \times (4/3)^2$. Load/input facewidth and diameter increase.

 iii. $120 \times (4/3)^3$. Facewidth also increases.

 iv. $120 \times (1.2)^2$. Square improvement due to material.

If root bending stresses control, the limits are:

   i. $120 \times (4/3)$. Increased tooth size.

  ii. $120 \times (4/3)^2$. Moment arm and radius of curvature increase.

 iii. $120 \times (4/3)^3$. Facewidth also.

 iv. $120 \times (1.2)$. Proportional to root material strength.

When tooth numbers are changed there is a small correction for tooth shape change but it is very small.

VII. INTERNAL GEARS AND RACKS

Internal gears may be treated in the same way as external gears with the use of a negative value for the radius of curvature. The taut string idea can still be used as indicated in Figure 2.6, but the path of contact (length of action) does not lie between the base circle tangent points. The limits of contact are still determined by the tip radii of the gears. The length of contact from the pitch point P to the internal gear limit is now governed by the radii of the pitch circle r_p and the tip circle $(r_p - a)$ so that

$$(r_p - a)^2 = r^2_p + PC^2 - 2 \, r_p \, PC \sin \phi$$

Contact stresses are low with internal gears since conformity is good; the negative radius of curvature of the internal gear gives an effective radius of curvature greater than that of the mating external gear. The low contact stresses allow internal gears to be made of material with lower surface hardness than external gears. Root stresses are also low since internal teeth are broad at the root. An internal and external gear of the same size will mate exactly with very low stresses and this principle is used in gear tooth couplings to give very compact flexible

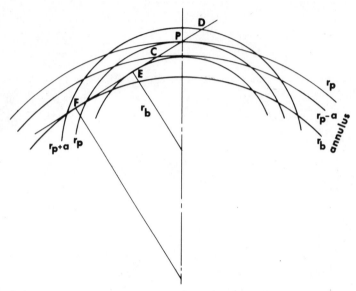

Figure 2.6. Internal gear contact. Contact is limited at D by the tip radius of the external gear and at C by the tip radius of the annulus teeth. EF is the "taut string."

drives. The mating external gear is "crowned" to allow misalignment of the gear axes.

Interference may occur between an internal and external gear if the external gear is not smaller than about 0.6 of the internal gear. This occurs because the tip of the external gear travels tangentially faster than the root of the internal gear. This effect is most likely to give trouble if small annuli are being machined.

Racks are a very special case of gears whose radius is infinite and whose "involutes" have become straight lines; unwrapping string ideas are of little help since the gear "center" and the base circle have gone to infinity. Racks are used principally to convert linear to rotary motion in designs where very high stiffness is needed. High forces are usually obtained more economically by hydraulics and, if stiffness is not important, wire-on-pulley designs are light and cheap but if, say, 20 tons of force is required for a 10 meter stroke with high rigidity as in a large broaching machine, there are no effective competitors to a rack and pinion drive.

Although the "base circle" position is unknown the pitch line (which corresponds to the pitch circle) is easily located since it is tangential to the mating gear pitch circle whose diameter is fixed by equating rack pitch to pinion circumferential pitch. The path of contact (length of action) is easily found from the approach contact governed by rack addendum (normally equal to m cosec ϕ) and the recess contact path which is governed by pinion addendum as in Section III.

Racks are relatively little used but the ideas of mating racks with gears are much used; gear shapes and corrections are usually defined in relation to the theoretical mating rack and the most common manufacturing method of hobbing is effectively form generation from the mating rack. Gear planing machines generate tooth form using racks which can be manufactured more easily and cheaply than hobbing or shaping cutters since they can be made on a simple surface grinder.

Contact between a pinion and rack differs from that between two gears in that the position of the pitch point relative to the pinion is unaltered by moving center distance and it is not possible to alter pressure angle by moving the pinion center.

VIII. CONTACT SLIDING

In Chapter 1 the relative sliding velocity at a contact was determined as $(\omega_1 + \omega_2)$PX where PX is the distance of the contact from the pitch point. The fractional loss of power due to friction with coefficient μ at the tooth contact is then

$$\frac{\mu \; \text{Force} \, (\omega_1 + \omega_2)\text{PX}}{\text{Force} \; \omega_2 \; r_b} \; \text{which is} \; \frac{\mu \, \text{PX}}{\cos \phi} \left(\frac{1}{r_1} + \frac{1}{r_2} \right)$$

In the very simple case where the contact ratio is 2, the average value of PX is $0.5 \, \pi \, m \cos \phi$ so that the average frictional losses are

$$\frac{\mu \; \pi \, m \cos \phi}{2 \cos \phi} \left(\frac{2}{N_1 m} + \frac{2}{N_2 m} \right) \; \text{which is} \; \pi \, \mu (1/N_1 + 1/N_2)$$

For normal gears, even with a coefficient of friction as high as 0.05, the losses are extremely small; two 40 tooth gears will give 0.75% losses. In practice losses are extremely small and gearbox losses are rarely controlled by friction at the teeth but are dominated by seal and bearing friction and windage and pumping losses.

The sliding velocity at contact is only relevant when a measure of the heat input into the oil film is required for oil breakdown estimates or when plastic gears are used without lubrication since wear rates are then dependent on the product of pressure and velocity.

Chapter 3

HELICAL GEARS

I. REASONS FOR USING HELICAL GEARS

The disadvantages of spur gears arise mainly from their vibration generation; any variation in the involute profile whether due to design, manufacturing or deflections will occur across the whole tooth face at the same time. This gives rise to a regular once-per-tooth excitation which can be, and often is, very powerful. The resulting vibration leads to heavy loads on the gears as well as to noise. A further disadvantage is that the extra strength available due to having two pairs of teeth in contact part of the time cannot be used since stressing is limited by that part of the cycle where only one contact is occurring.

Helical gears can be considered as a pack of thin spur discs staggered so that contact is at a different part of the profile for each of the discs. This has the effect of averaging out the errors on each individual disc; this averaging effect is very powerful due to the elasticity of the teeth and gives the result that teeth which are only correct in profile to within 10 microns can average errors to run smoothly under load within 1 micron. There is an additional bonus on strength because at any instant in time about half the discs will have double contact if the contact ratio is about 1.5 so that stressing can be based on roughly 1.5 times the facewidth instead of the facewidth.

Manufacture and mounting of packs of discs would be difficult and expensive so the gear is made continuous with teeth which follow a helix on the gear. This gives rise to axial forces which can be troublesome, unlike spur gears, but the vibration and strength advantages far outweigh the disadvantages of axial thrust and slightly increased manufacturing cost.

II. DEFINITIONS AND BASIC GEOMETRY

A simple involute is described when a string unwinds from a base disc; in helical gears the same concept is used, i.e., a band unwinds from a base cylinder (one gear) and winds onto another base cylinder. The essential difference is that the line on the unwrapping band is no longer square on the band (generating a spur)

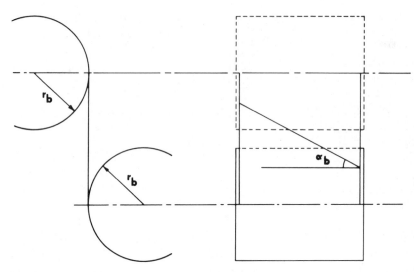

Figure 3.1. Generation of "involute helicoid." A line set at an angle α_b on the band which unwraps from one base cylinder will generate a tooth flank of a helical gear.

but has been set at an angle on the band. Perforations set drunkenly on a toilet roll would generate "involute helicoid" surfaces as they unroll.

Figure 3.1 shows an unwrapping band between two base cylinders and we define the line inclination from a line parallel to the axes of rotation as α_b the base helix angle. It cannot be measured directly despite its importance in design and so in practice it is best to specify the "lead" of a gear L. This is the distance it is necessary to travel axially along a helical gear for a given tooth to rotate through 360°. The relationship between L and α_b is given by

$$\tan \alpha_b = 2\pi r_b/L \quad \text{so} \quad L = 2\pi r_b \cot \alpha_b$$

Two mating gears must have the same α_b, though of different hands, but will have different leads in the ratio of the numbers of teeth.

The other radius of importance on a gear is the pitch radius (or to be more precise the reference radius) and the helix angle α_p can be measured at this radius. The relationship with lead still applies so that

$$L = 2\pi r_b \cot \alpha_b = 2\pi r_p \cot \alpha_p \quad \text{and since} \quad r_b = r_p \cos \phi_t$$

$$\tan \alpha_p = \tan \alpha_b \sec \phi_t$$

The suffix t indicates a measurement in the *transverse* plane, defined as a plane perpendicular to the axis of rotation so that the pressure angle ϕ_t is the pressure angle as seen when looking at the end face of a helical gear.

Although the fundamental involute shape and mating action occurs in the transverse plane we usually specify the tooth pressure angle and size by looking in the *normal* plane. This is a plane which is perpendicular to the helix at the pitch point on a particular tooth; it is a different plane for each tooth round a gear. The reason for the interest in the normal plane is that most manufacturing processes operate along the helix and so generate a standard tooth form in the normal plane. Pedantically, a tooth form generated from an involute cutter or rack in the normal plane is not a correct involute in the transverse plane as it should be, but the difference is negligible and is much less than design and manufacturing errors.

Since the normal plane is of interest we define pitch, module, and pressure angle in the normal plane and denote them with the subscript n to show that they are not measured in the transverse plane.

Conventional labelling uses subscript n for the normal plane and no subscript for the transverse plane; it is however perhaps wiser to use subscript t for the transverse plane since there is less danger of careless mistakes being made. In the same way, conventionally, base radii, pitch, and helix angles are labelled but pitch radii, pitch, and helix angles are not; again, it is safer to label "p" for clarity and this convention will be followed in this text.

Looking at the pitch cylinder as shown unwrapped in Figure 3.2 the lines represent successive teeth; the distance apart of the lines is the normal pitch p_n and is equal to $\pi\, m_n$. Transverse pitch p_t is, as with a spur gear, $\pi\, m_t$ and the other pitch of interest is the axial pitch p_a, which is $p_t \cot \alpha_p$; axial pitch is also equal to the lead divided by the number of teeth on the gear. p_t is $p_n \sec \alpha_p$ so then $m_t = m_n \sec \alpha_p$.

The advantage of specifying normal module is to simplify manufacturing since only standard module hobbing cutters need be used. The helix angle can

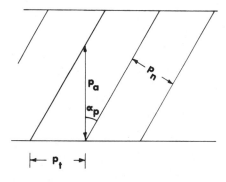

Figure 3.2.   View of unwrapped pitch cylinder. The lines represent the positions of pitch radius on successive teeth. The corresponding inclination is the helix angle at pitch radius.

then be chosen to give the required center distance for a particular design. A simple example is:

> A pair of helical gears is to operate with a center distance of 87 mm and a ratio of 2.25 to 1, using standard 2 mm module teeth. What is the helix angle?

Since the ratio is 9:4 the total number of teeth must be divisible by 13 and since the transverse module is $m_n \sec \alpha_p$ then

$$(N_1 + N_2) \times 0.5 \times m_n \sec \alpha_p = 87 \text{ mm.}$$

$7 \times (4 + 9)$ is > 87 so $6 \times (4 + 9)$ teeth are required, i.e., 24 and 54 teeth and $\sec \alpha_p = 87/78$ so $\alpha_p = 26.29°$.

III. TRANSVERSE PRESSURE ANGLE

The normal pressure angle is standardized (at $20°$) but it is the transverse pressure angle which controls contact ratio so we require a relationship between ϕ_n and ϕ_t.

Figure 3.3 shows a small section of surface at the pitch radius on a gear flank (or the surface of the theoretical mating rack). PB is part of the line on the unwrapping band which generates the tooth surface and is the intersection of the

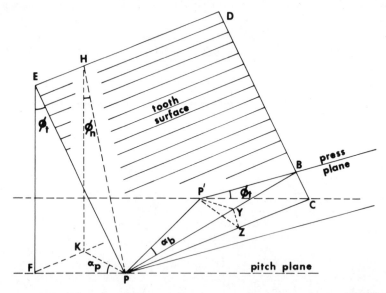

Figure 3.3. Relationship between normal and transverse pressure angles. The pressure plane (unwrapping band) meets the pitch cylinder (a plane if a very small section is taken) in the line PP′. A small section of the tooth surface is represented by EDCP.

pressure plane (i.e., the band) with the tooth surface; PP' is in the axial direction. Triangle EFP lies in the transverse plane and HKP is in the normal plane.

FEHK is a rectangle where $FP = EF \tan \phi_t$, $KP = HK \tan \phi_n$, $KP = FP \cos \alpha_p$, and $EF = HK$ so

$$\tan \phi_n = \tan \phi_t \cos \alpha_p$$

Also if $P'YZ$ is another section perpendicular to PC so that $P'Y$ is perpendicular to PB, a line of contact, and normal to the tooth surface PCDE and at an angle ϕ_n to the pitch plane then $P'Z = PP' \sin \alpha_p$, $P'Y = PP' \sin \alpha_b = P'Z \cos \phi_n$ giving $\sin \alpha_b = \sin \alpha_p \cos \phi_n$ and we can also substitute $\tan \alpha_p = \tan \alpha_b \sec \phi_t$ from the lead relationship so that $\cos \alpha_p \cos \phi_n = \cos \phi_t \cos \alpha_b$.

Collecting together the relevant geometric relationships gives

$$
\begin{array}{ll}
r_b = r_p \cos \phi_t & 2r_p = Nm_t \\
p_t = \pi m_t & p_b = \pi m_t \cos \phi_t \\
m_n = m_t \cos \alpha_p & p_a = p_t \cot \alpha_p
\end{array}
$$

$$
\begin{aligned}
\tan \alpha_b &= \tan \alpha_p \cos \phi_t \\
\tan \phi_n &= \tan \phi_t \cos \alpha_p \\
\sin \alpha_b &= \sin \alpha_p \cos \phi_n \\
\cos \alpha_p \cos \phi_n &= \cos \phi_t \cos \alpha_b.
\end{aligned}
$$

Returning to the example at the end of Section II, we can now calculate the transverse pressure angle and the transverse contact ratio for standard teeth.

$\tan \phi_n = \tan \phi_t \cos \alpha_p$ so with $\phi_n = 20°$ and $\alpha_p = 26.29°$, $\phi_t = 22.1°$

The pitch radii are $87 \times 4/13$ and $87 \times 9/13$ and for both gears the addendum will be 2.0 mm; note that the addendum is the *normal* module, not the transverse module for standard teeth. Hence the approach distance is given by

$$28.769^2 = 26.769^2 + 2 \times 26.769 \, x_1 \sin 22.1° + x_1{}^2$$

so $x_1 = 4.506$ and similarly $x_2 = 4.879$

The base pitch is $2.0 \times \sec 26.29° \times \cos 22.1° \times \pi = 6.493$ so the contact ratio is 1.445.

IV. LENGTH OF LINES OF CONTACT

The total length of the lines of contact is important because it controls permissible loadings on a pair of gears and any variation in the length of the lines of contact gives contact stiffness variation which increases vibration levels.

If, in Figure 3.4, we look at the pressure plane, i.e., the unwrapping band we see a series of angled lines which are lines of contact on tooth flanks moving from one base cylinder to the other. Although our hypothetical band extends

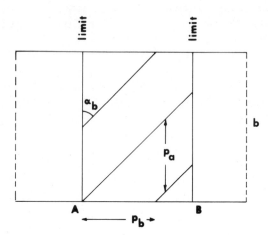

Figure 3.4. Length of lines of contact. View of pressure plane containing the lines which are limited at A and B by loss of contact at the teeth tips. b is the facewidth of the gears.

from base cylinder to base cylinder, contact can only occur with a central area bounded by the contact limits where the pressure plane reaches the tooth tips as with spur gears.

The distance AB between the contact limits is calculated as in Chapter 2 Section III and is $r_c \times p_b$ or $r_c \times \pi \, m_t \cos \phi_t$ where r_c is the contact ratio. The length of a facewidth of contact line is $b \sec \alpha_b$ and there are on average r_c contact lines so the average length of line of contact is $r_c \, b \sec \alpha_b$. In the particular case when the facewidth is an exact number of axial pitches, as shown in Figure 3.4, the length of the line of contact does not vary and is $r_c \, b \sec \alpha_b$ but in the

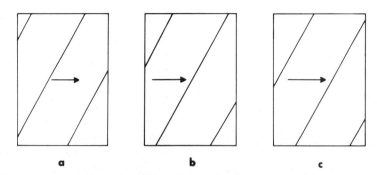

Figure 3.5. Pressure plane views. successive positions of the contact lines as meshing occurs are shown. (a) is a condition of increasing line length, (b) is a maximum and (c) shows a shortening condition.

more general case where the gear has been badly designed and the facewidth is not an exact number of axial pitches, the variation in contact must be calculated.

Looking at Figure 3.5, a view of the part of the pressure plane, which contains the lines of contact, if the "unwrapping band" is moving from left to right we will see the set of contact lines moving to the right in the window. As shown in (a) there are two lines lengthening and one shortening so the total line length is increasing whereas in (b) one is shortening and one is lengthening so the line length is not changing; correspondingly in (c) two are shortening so the total length is reducing.

An example of calculation of line length is illustrated in the following.

Two identical gears with $\phi_n = 20°$ mesh; the base helix angle is $30°$ and the facewidth is 7 times the normal module. What are the maximum, minimum, and mean values of the length of line of contact if the contact ratio is 1.6?

With α_b of $30°$ and ϕ_n of $20°$ the pitch helix angle α_p is given by $\sin \alpha_b = \sin \alpha_p \cos \phi_n$ so $\alpha_p = 32.1°$. The transverse module is then $m_t = m_n \sec \alpha_p$ and the base pitch is $\pi m_t \cos \phi_t$; to get ϕ_t use the relation $\cos \phi_t \cos \alpha_b = \cos \alpha_p \cos \phi_n$ so that finally the base pitch is $\pi m_n \sec \alpha_p . \cos \alpha_p \cos \phi_n / \cos \alpha_b$ which comes out as $3.41 m_n$.

Sketching a view of the pressure plane, see Figure 3.6(a), gives the dimensions shown and the particular view is for the condition where the total line length is just about to start decreasing. The length of line shown is then $7 \sec 30° + (5.45 - 3.41) \operatorname{cosec} 30° + (7 \tan 30° - 3.41) \operatorname{cosec} 30°$ which is $13.45 m_n$.

Figure 3.6(b) is correspondingly for a minimum condition and gives the values of $3.41 \operatorname{cosec} 30° + 7 \sec 30° - (2 \times 3.41 - 5.45) \times \operatorname{cosec} 30° = 12.17 m_n$.

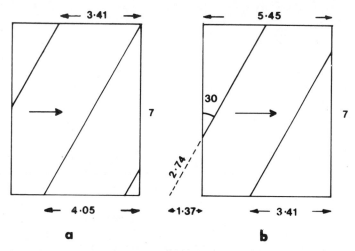

Figure 3.6. Pressure plane for example: (a) is a maximum length condition and (b) is a minimum. All dimensions are multiples of the normal module m_n.

The average value of line length is $7 \times 1.6 \times \sec 30° = 12.92 \, m_n$ which is not the arithmetic mean of the maximum and minimum. The expressions for the line lengths can be simplified in particular cases but it is necessary to determine first how many full axial pitches are involved. The effort of doing these maximum and minimum length of line calculations is usually not justified since a design should always have an exact number of axial pitches for smooth running.

When carrying out gear design it is necessary to have high accuracies for calculating factors such as helix angles because these should be quoted accurate to one part in 10^5 on medium size gears (1 micron on 100 mm facewidth) and so calculation should be an order better, i.e., to 7 figures. However it should then be remembered that lengths of contact lines, etc., are unlikely to be within 10 microns and so should never be quoted to less than one micron.

V. SURFACE CURVATURE

Hertzian contact stresses are affected by contact radii so it is advisable to be able to estimate the effects of design changes, e.g., of helix angle.

The force between the teeth is in the normal plane so it is the curvature in this plane which is important. The conventional way of determining the surface curvature is to use the concept of a "virtual spur" wheel which is defined as the spur gear which would approximate to the shape of the helical gear in the region of contact.

Taking a normal section of a particular tooth would give a section (as shown in Figure 3.7) of the pitch cylinder which is an ellipse of minor axis r_p and major axis $r_b \sec \alpha_p$. The radius of curvature at the tooth of interest is $r_p^2 \sec^2 \alpha_p / r_p$ which is $r_p \sec^2 \alpha_p$ and the tooth pitch is $\pi \, m_t \cos \alpha_p$ since $m_n = m_t \cos \alpha_p$. This gives an equivalent spur wheel with radius increased by $\sec^2 \alpha_p$ and tooth pitch reduced by $\cos \alpha_p$ compared with a transverse section of the original

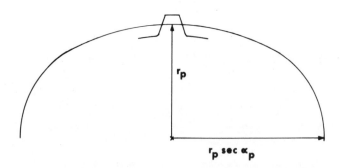

Figure 3.7.   Virtual spur gear. Cross section of a gear perpendicular to the pitch helix on a gear of pitch radius r_p.

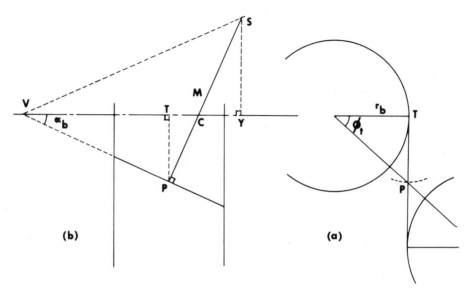

Figure 3.8. Gregory method. Geometry of tooth surface curvature by finding the curvature of the on which contact points lie. V is the cone vertex and P the pitch point. (a) is and edge view and (b) a true view of the pressure plane.

helical gear so the equivalent number of teeth is increased by $\sec^3 \alpha_p$. Calculations of stress in the normal plane then proceed as for a spur gear with $N \sec^3 \alpha_p$ teeth, pressure angle ϕ_n, and module m_n. The radius of curvature at the pitch is given by $0.5 \times N \sec^3 \alpha_p \times m_n \times \sin \phi_n$.

Although use of the equivalent "virtual spur" gear is the normal method it is not quite correct and a rather more exact and more elegant method was devised by Dr. R. W. Gregory.

Figure 3.8(a) shows a transverse section of a pair of gears (meshing) at the pitch point and Figure 3.8(b) gives the corresponding view on the pressure plane projected from Figure 3.8(a). In the *transverse* section all teeth flanks have their center of curvature located at T and so all the working surface of the tooth lies on a conical surface whose axis is VT and which has a semi-angle α_b. V is located as the point where a line through T parallel to the axis of rotation, i.e., the center line of the cone, meets the line of contact which passes through P.

We now require the radius of curvature of the conical surface perpendicular to PV. This may be obtained either by determining the semi-axes of the ellipse section which appears as PS or by invoking the theorem which states that the center of principal curvature of any surface of revolution lies on the axis of revolution, i.e., at C. Using the ellipse method we have PT as $r_b \tan \phi_t$ and if we call SV ℓ the major axis of the ellipse is $\ell \sin 2\alpha_b/2$. The minor axis of the ellipse is given by the height of M, on the cone surface midway between P and S above the pressure plane. The point M lies on a transverse section of the cone midway between Y and T so the section radius is 0.5 ($\ell \cos \alpha_b + \ell \cos 2\alpha_b \cos \alpha_b$) tan α_b which simplifies to $\ell \sin \alpha_b \cos^2 \alpha_b$. The distance of M from VT in view (b) is ($\sin 2\alpha_b \ell/2 - \ell \cos 2\alpha_b \tan \alpha_b$) $\cos \alpha_b$ which simplifies to $\ell \sin \alpha_b \sin^2 \alpha_b$. Squaring and subtracting, Pythagoras gives the square of the height of M as $\ell^2 \sin^2 \alpha_b \cdot \cos 2\alpha_b$ and so the curvature of the ellipse, b^2 / a is $2\ell^2 \sin^2 \alpha_b \cos 2\alpha_b / \ell \sin 2\alpha_b$ which reduces to $\ell \tan \alpha_b \cos 2\alpha$ which is the distance PC and, hence, is $r_b \tan \phi_t \sec \alpha_b$.

The answers obtained by the two methods are slightly different, $r_b \tan \phi_t \sec \alpha_b$ by the Gregory method compared with the equivalent spur method which gives $r_p \sec^2 \alpha_p \sin \phi_n$. The ratio of the two expressions is $\cos^2 \phi_t / \cos^2 \phi_n$ which is slightly less than 1.

The reason for the difference in the end answers is that the virtual spur method determines the curvature in the normal plane perpendicular to the pitch helix, i.e., as if looking straight along a tooth whereas the Gregory method gives the curvature looking correctly along the line of contact which is slanting up the face of the tooth.

The Gregory method also shows clearly how the radius of curvature varies along the length of the tooth as the distance PC varies; the expression obtained is for the curvature at the pitch point.

VI. TORQUE COMPARISONS

Helical gears follow the same rules as spur gears for variation in size since similarly

Torque α b $\times$ (Nm) $\times$ m $\times$ σ for root stresses and
Torque α b $\times$ (Nm)^2 $\times$ σ^2 for contact stresses

The other possible variable is helix angle. If we take P as the normal force between the gears then when root bending stresses limit.

Torque α P $\cos \alpha_b \cdot r_p \cos \phi_t$

Provided that the module m_n remains the same so that the tooth strength remains the same then P α m_n $\times$ σ $\times$ length of line of contact. So

Torque α m_n $\times$ σ $\times$ ($br_c \sec\alpha_b$). $\cos \alpha_b \cdot r_p \cos \phi_t$
i.e., α b $\times$ (r_p) $\times$ m_n $\times$ $\sigma r_c \cos \phi_t$

To a first order the effects of helix angle cancel out but as helix angle increases r_c decreases and ϕ_t increases, both giving a slight reduction in overall torque: there is a slight compensating increase in basic tooth strength as the number of teeth in the equivalent spur wheel increases. It should also be noted that increasing helix angle but keeping pitch diameter constant reduces the number of teeth. When root strength is the limiting factor the strongest gear (theoretically, taking no account of defects, vibrations, etc.) is obtained by using the minimum helix angle consistent with smooth running, i.e., such that the facewidth is one axial pitch and having the minimum number of teeth to give maximum tooth size.

When contact stresses limit, as is more common, the effects are slightly different. With P as the total normal force between teeth

$$\text{Torque} \propto P \cos \alpha_b \; r_p \cos \phi_t$$

and $P \propto (b \, r_c \sec \alpha_b) \times \sigma^2 \times R_e$ the radius of curvature.

The effective radius of curvature is $r_b \tan \phi_t \sec \alpha_b$ at the pitch point so

$$\text{Torque} \propto \cos \alpha_b \; r_p \cos \phi_t \, b \, r_c \sec \alpha_b \times \sigma^2 \times r_b \tan \phi_t \sec \alpha_b, \text{ i.e.,}$$
$$\propto b \, r_p \sin \phi_t \sec \alpha_b \, r_c \, \sigma^2 \, r_b$$
$$\text{or } b \, r_p^2 \, \sigma^2 \, r_c \sin^2 \phi_t \sec \alpha_b$$

$r_c \times \sin \phi_t$ shows little variation with helix angle so the torque varies as sec $\alpha_b \times \sin \phi_t$, both of which increase with helix angle so that permissible torque rises slightly with helix angle when contact stresses limit. The number of teeth is not important for contact stresses but must reduce or a smaller module must be used as helix angle rises. As with root stresses the assumption is that the facewidth is an integral number of axial pitches.

The overall conclusion is that there is little point in altering helix angle in an attempt to influence gear strength and the basic rules remain that torque is proportional to volume when contact stress limits and to plan area ($b \times r_p$) times module when root stresses limit.

Chapter 4

OTHER GEAR DRIVES

I. SKEW GEARS

Drives between axes which are not parallel present many problems if they have to withstand even moderate loads. Two helical gears will mesh together to give a skew drive, e.g., two 15° right hand helix gears will give a 30° angle drive but since contact is only point contact instead of line contact, loads are restricted. The effect of the crossed axes is to superimpose high relative sideways velocities at the point of contact, much higher than the normal parallel-axis sliding velocities and this tend to increase the likelihood of oil breakdown and scuffing.

Bevel gears may be used to drive between axes which intersect, the most usual requirement being for axes which are perpendicular. Straight toothed bevel gears are the conical equivalent to cylindrical spur gears and operate in the same way but are much more sensitive to assembly errors since the distances of the gears from the point of intersection of the axes must be maintained accurately. Straight bevels are rarely made accurately and can only be checked on transmission error equipment so they should never be used where high speeds or accuracies are required.

Spiral bevel gears, the conical equivalent of helical gears give a smoother drive than straight bevels but are still relatively "rough." The further development of hypoid bevels for use in cars came from the requirement to drop the center line of the propellor shaft to reduce tunnel height in conventional rear wheel drive cars. The axes of hypoids are at 90° but are offset by typically 50 mm giving a great gain in rear seat foot space; the penalty lies in increased contact sliding velocities.

The geometry of the hypoid bevels is extremely complex and is best left to specialists; gears made specially are extremely expensive due to the very high costs of setting up and so designs which require right angle drives must be avoided unless it is possible to use a standard car or truck rear axle unit. Standard units are mass-produced and, hence, with immense development expertise are extremely cheap and surprisingly accurate; typically a well made back axle transmitting 100 kW at 5000 rpm will drive accurately within 5 μm at tooth frequency, as accurately as any gear which is not a ground helical parallel axis gear.

Worms and wheels have the advantage of being capable of giving high reduction ratios very compactly but have high sliding velocities. It is usual to make one member, the wheel, in a material such as phosphor bronze which will tolerate poor lubrication conditions; this is because it is difficult to design and to manufacture sufficiently accurately that in operation, under load, there is a correctly converging oil film between the mating surfaces. Poor lubrication conditions associated with diverging oil films will cause rapid local wear until the surface has reached a satisfactory shape; unfortunately change of temperature or load may then require a different shape so further wear occurs.

Bevels and worms and wheels all suffer from a lack of tolerance of position errors unlike involute parallel axis gears and so must be mounted extremely carefully. Even a perfectly cut pair of gears will give large errors at once per tooth frequency if the gears are mounted with alignment or distance errors. Only transmission error testing (see later) is of use and it is advisable to have some method of adjusting axes in situ if high accuracy is required. Where accuracy of drive is required, it is usually essential to have bevel drives on high speed shafts only and then reduce speed with helical gears so that the inaccuracies of the bevel gears are scaled down. Worms and wheels can be accurate but extreme care is needed.

II. WILDHABER-NOVIKOV GEARS

The most common type of gear damage occurs as pitting of the working surface due to excessive local Hertzian pressures. Materials cannot be further improved so it is necessary to investigate alternative methods of improving highly loaded gears such as increasing the effective radius of curvature at the contact.

Involute gears can only achieve this by increasing pressure angle which, unfortunately, reduces contact ratio and slightly increases transmitted load. The opposite approach of reducing pressure angle gives increased contact ratio and, hence, decreases load per unit facewidth on each tooth but decreases radius of curvature and gives a more slender and, hence, weaker tooth shape.

In an attempt to reduce contact stresses Wildhaber-Novikov gears use highly conforming profiles which are both arcs of circles but have one convex and one concave as shown in Figure 4.1. The high conformity gives low contact stress and the relatively low, thick teeth are strong in bending. A narrow gear will not give a smooth drive however as contact is intended to occur only at the pitch point P. To achieve a smooth drive helical gears must be used and the contact between the gears is then in theory a series of ellipses spaced at the pitch line across the width of the gear as shown in Figure 4.2. As the gears rotate contact occurs only at the pitch radii on the line joining the gear centers but the contact ellipses travel axially along the gear; in practice the "ellipses" tend to be surprisingly large and are curved. Despite the fact that the contact is a series of

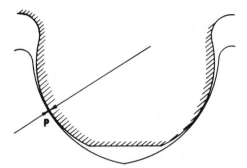

Figure 4.1. Tooth shapes for Wildhaber-Novikov gears.

"points" instead of a "line" contact as with involutes, the total area of contact is larger due to the high conformity and surface stresses are reduced.

As the gears give low contact stresses and have squat strong teeth it would appear that they are preferable to involute gears and for some relatively rough low-speed applications they have marginal advantages. The problems encountered with them are:

i. The gears are sensitive to center distance so that gear diameters and case center distances must be tightly toleranced.
ii. The squat teeth are extremely rigid. This does not matter at very low speeds but at moderate speeds the forces developed at the teeth due to

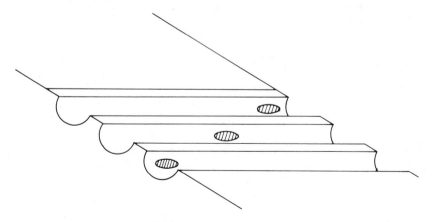

Figure 4.2. Sketch of contact ellipses on a Wildhaber-Novikov gear. The ellipses shown are roughly in the plane containing the center lines of the mating gears.

manufacturing errors are much higher than with involutes and give very high vibration levels.

iii. The tooth surfaces cannot be generated and so form can only be ground by a contoured wheel or hobbed. This manufacturing restriction limits use in large sizes and inclines this type of gearing to the softer materials which do not require grinding.

iv. The direction of the force between teeth remains constant and, if very high accuracy is maintained, the magnitude of the force remains constant but the position of the force varies along the gear, i.e., axially as the contact ellipses disappear off one end of the face width and corresponding contacts start at the other end. This variation in force position is an inherent cause of vibration and noise generation and cannot be eliminated in practice; it can be greatly reduced by silhouetting but this reduces load capacity (1).

The combination of sensitivity to manufacturing errors, inherent high vibration levels, and problems in manufacture have effectively eliminated circular arc gears as competitors to the involute since it is normally possible to solve a problem more economically with conventional gears.

III. GEAR PUMPS

Two similar involute spur gears mounted inside a close fitting case as shown in Figure 4.3 will act as a pump or in reverse as a motor. The central mesh between

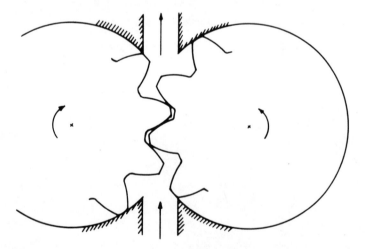

Figure 4.3.   Gear pump. Oil cannot flow back through the mesh but is carried from low to high pressure region in the spaces between the teeth. This will also act as a motor.

the gears acts both as a drive to transmit rotation between the gears and as a seal to prevent leakage of oil from high pressure (at the top) to low pressure. Low pressure oil is carried round the gear in the gaps between the teeth and the volume delivery per revolution corresponds to the tooth space volume.

Leakage must be kept low so clearance between tooth tips and the casing and between gear and faces and the casing must be kept low; backlash between the gears must also be minimized. This form of pump is cheap and extremely robust but does not have a high efficiency since narrow gaps with high sliding velocities give either high viscous drag losses or high leakage losses.

Problems of noise and vibration occur when the low pressure oil between the teeth suddenly rises in pressure. The bulk modulus of oil is 1.4×10^9 (2×10^5 lb/sq. in.) so that a typical delivery pressure of $7 \times 10^6 \mathrm{N/m^2}$ (1000 lb/sq. in.) gives 0.5% volume decrease and transmits pressure pulses through the system. Problems can also occur at the mesh point if oil is trapped in the tooth roots and subjected to volume changes.

In a given pump diameter, increase of delivery per revolution can only be obtained by increasing the volume between the teeth and, hence, decreasing the number of teeth. An extreme case is the Roots blower type of pump, usually used for gases where the "gears" have only 3 teeth and, hence, have a large swept volume for revolution. The gears conform and pump relatively large volumes

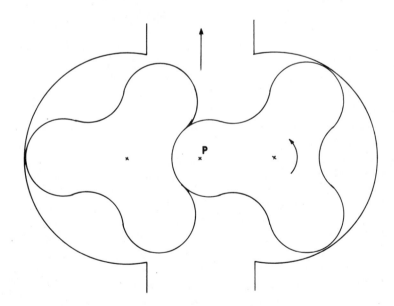

Figure 4.4. Roots blower. The rotors conform but cannot drive so external gears are needed.

but, although the common normal at the point of contact always passes through the pitch point P, see Figure 4.4, the direction of the common normal may be along the line joining centers and so the "gears" will not drive. It is necessary to have additional external gears to keep the rotors synchronized. As with oil pumps, compressibility effects give relatively large pressure fluctuations, but the Lysholm compressors which are the helical derivatives of Roots blowers over-come this by gradual compression with axial inlet and outlet.

IV. GEAR TOOTH COUPLINGS

Gears are used for couplings to transmit high torque and accommodate misalign-ment and axial movements. Their tolerance of large axial movements and of radial movement allows them to be smaller and lighter than rival diaphragm couplings giving whirling advantages but they may wear and can occasionally generate vibration.

Each end of the coupling is formed by having a crowned spur gear inside an annulus which has teeth of the same size. The clearance between gear and annulus is small and the conformity between the mating surfaces is good so that contact pressures are nominally very low; in practice manufacturing errors pro-duce high local stresses. Misalignments of several degrees can be accommodated as the crowning allows movements of several millimetres axially at the teeth. The gear is self-centering under torque and it is bad practice to have a very low clear-ance between gear and annulus as this can lead to lockup and resulting failure under some temperature transient conditions (2).

In operation, misalignment produces a bending moment due to friction at the gear teeth. This moment is not high but occasionally can be significant in stressing as it has a fatigue effect unlike drive torques which are steady in ampli-tude. Making the assumption that the torque loads are taken evenly round the coupling gives a bending moment of μ T sec $20°.2/\pi$ where T is the transmitted torque and μ is the coefficient of friction at the teeth. The value of μ is open to debate as the gear coupling is usually grease lubricated but has relatively rough surfaces and insufficient movement to generate hydrodynamic lift. A value of about 0.15 leads to a shaft bending moment of about 10% of the transmitted torque; this can load retaining nuts heavily.

When driver and driven shafts are parallel but offset, there is a lateral force needed for equilibrium equal to 4 μT sec $20°/\pi\ell$ where ℓ is the length of the coupling. This lateral force overhung a large distance from a small diameter bearing will give shaft fatigue bending stresses which are of the same order of size as the drive torque shear stresses.

Heavy misalignment will give some heat generation in the coupling. The power is 4 n μT sec $20°.\phi$ where ϕ is the angular misalignment in radians and n is the number of revolutions per second. Some extreme figures of 6000 rpm, 0.01

radians, 2000 Nm torque, μ = 0.2 give a power of 1600 watts; this is negligible compared with the transmitted power of 1256 kW but will raise local temperatures. The conductivity of steel is low, 54 W/mK (30 Btu/sq. ft./hr./°F/ft.) so any heat generated does not travel far.

The other critical condition with a gear tooth coupling occurs when there is no misalignment. It is sometimes wrongly argued that this cannot occur but what can and does occur is that small misalignments can easily be accommodated by elastic bending of shafts and the coupling does not then slip. When this happens the axial forces that can be generated with thermal expansion of a turbine are equal to μT sec 20°/r where r is the pitch circle radius. 2000 Nm torque with 0.1 m radius and μ = 0.2 gives a possible axial force of 4000 N (1000 lbf) which must be taken by the turbine thrust bearing.

REFERENCES

1. Rouverol W. S., Constant Tooth Load Gearing. World Congress on Gearing. Paris, June 1977, pp. 257–268. (Also Rolling Contact Gear Co., Berkeley, California.)
2. Smith J. D., Tooth Gear Coupling Problems. Inst. Mech. Mechanisms 76 – Mechanical Drive Systems UMIST, September 1976, Paper 7.

Chapter 5

MANUFACTURING METHODS

I. PLANING

The shape of the space between gear teeth is complex and varies with the number of teeth on the gear as well as tooth module, so most gear manufacturing methods generate the tooth flank instead of forming. Either a rack section may be used, linked so that it moves exactly a pitch for each gear tooth or another involute in the form of a circular cutter with cutting "gear teeth" which are capable of generating a mating shape.

Planing uses a reciprocating rack, stroking in the direction of the helix on a gear with a gradual generation of form as the rack effectively rolls round the gear blank. The rack is relieved out of contact for the return stroke as in normal shaping or planing. It has the great advantage that the cutting tool is a simple rack with (nearly) straight sided teeth which can easily be ground accurately. After several tooth spaces have been generated the rack reaches the end of its travel and is traversed back an exact number of pitches ready to generate the next few teeth. This method is little used for high production because it is relatively slow in operation due to the high tool and slide mass; for jobbing purposes the slow stroking rate does not matter and low tool costs give an advantage where unusual sizes or profile modifications are required.

Although planing cannot normally compete with hobbing it has the advantage that it requires little or no runout or clearance at the far side of a gear so that gears may be planed close to larger diameter shoulders or double helical gears may be cut without having to leave a clearance gap between the helices.

Alteration of helix angle is achieved by rotating the direction of stroking but the cutting rack does not usually have to be altered.

II. SHAPING

Shaping is inherently similar to planing but uses a circular cutter instead of a rack and the resulting reduction in the reciprocating inertia allows much higher stroking speeds; modern shapers cutting car gears can run at 2,000 cutting strokes per minute. The shape of the cutter is roughly the same as an involute gear but the tips of the teeth are rounded.

39

The generating drive between cutter and workpiece does not involve a rack or leadscrew since only circular motion is involved. The tool and workpiece move tangentially typically 0.5 mm for each stroke of the cutter. On the return stroke the cutter must be retracted about 1 mm to give clearance otherwise tool rub occurs on the backstroke and failure is rapid. The speed on this type of machine is limited by the rate at which some 50 kg of cutter and bearings can be moved a distance of 1 mm. The accelerations involved require forces of the order of 5000 N (0.5 Ton) yet high accuracy must be maintained.

The advantages of shaping are that production rates are relatively high and that it is possible to cut right up to a shoulder. Unfortunately, for helical gears, a helical guide is required to impose a rotational motion on the stroking motion; such helical guides cannot be produced easily or cheaply so the method is only suitable for long runs with helical gears since special cutters and guides must be manufactured for each different helix angle. A great advantage of shaping is its ability to cut annular gears such as those required for large epicyclic drives.

When very high accuracy is of importance the inaccuracies in the shaping cutter matter since they may transfer to the cut gear. It is obvious that profile errors will transfer but it is less obvious than an eccentrically mounted or ground cutter will give a characteristic "dropped tooth." There are several causes for "dropped tooth" but it occurs most commonly when the diameter of the workpiece is about half, one and a half, two and a half, etc times the cutter diameter. If the cutter starts on a high point and finishes on a low point during the final finishing revolution of the gear the peak to peak eccentricity error in the cutter occurs between the last and first tooth of the final revolution of the cut gear; as the cumulative pitch error of the cutter may well be over 25 microns there is a sudden pitch error of this amount on the cut gear. The next gear cut on the machine may however be very good on adjacent pitch if the final cut happened to start in a favourable position on the cutter.

Various attempts have been made to prevent this effect, in particular by continuing rotation without any further cutter infeed but if the shaping machine is not very rigid and the cutter very sharp then no further cutting will occur and the error will not be removed.

III. HOBBING

Hobbing, the most used metal cutting method, uses the rack generating principle but avoids slow reciprocation by mounting many "racks" on a rotating cutter. The "racks" are displaced axially to form a gashed worm. The "racks" do not generate the correct involute shape for the whole length of the teeth since they are moving on a circular path and so the hob is fed slowly along the teeth either axially in normal hobbing or in the direction of the helix in "oblique" hobbing. Figure 5.1 shows the principle.

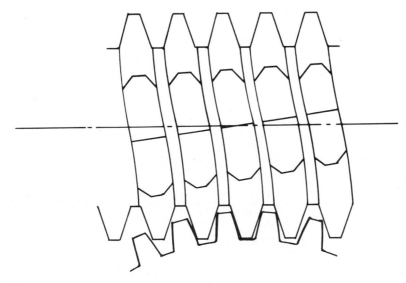

Figure 5.1. Sketch of hobbing process.

Metal removal rates are high since no reciprocation of hob or workpiece is required and so cutting speeds of 40 m/min (120 ft/in) can be used for conventional hobs and up to 150 m/min (450 ft/in) for carbide hobs. Typically with a 100 mm diameter hob the rotation speed will be 100 rpm and so a twenty tooth workpiece will rotate at 5 rpm. Each revolution of the workpiece will correspond to 0.75 mm feed (30 mil) so the hob will advance through the workpiece at about 4 mm (150 mil) per minute. For car production roughing multiple start hobs can be used with coarse feeds of 3 mm per revolution so that 100 rpm on the cutter, a two-start hob and a 20 tooth gear will give a feed rate of 30mm/minute.

The disadvantage of a coarse feed rate is that a clear marking is left on the workpiece, particularly in the root, showing a pattern at a spacing of the feed rate per revolution. This surface undulation is less marked on the flanks than in the root and is not important when there is a subsequent finishing operation such as shaving or grinding. When there are no further operations the feed per revolution must be restricted to keep the undulations below a limit which is usually dictated by lubrication conditions. The height of the undulations in the root of the gear is given (see Figure 5.2) by squaring the feed per revolution and dividing by four times the diameter of the hob; 1 mm feed and 100 mm diameter gives 2.5 micron (0.1 mil) high undulations in the root. On the gear flank (at $70°$ on a large gear) the undulation is roughly $\cos 70°$ as large, i.e., about 0.85 micron.

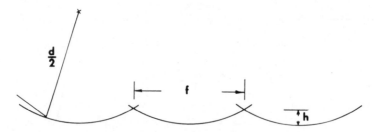

Figure 5.2.   Undulation heights when hobbing. f is the feed per revolution of
the hob, d is the hob diameter and h is the height of the undulation.

Accuracy of hobbing is normally high for pitch and for helix, provided
machines are maintained; involute is dependent solely on the accuracy of the
hob profile. As the involute form is generated by as many cuts as there are
gashes on the hob the involute is not exact, but if there are say 14 tangents generating a flank of 20 mm radius of curvature about 4 mm high the divergence
from a true involute is only about half a micron; hob manufacturing and mounting errors can be above 10 microns. Use of two-start hobs or oblique hobbing
gives increased error levels since hob errors of pitching transfer to the cut gear.

IV. BROACHING

Broaching is not usually used for helical gears but is useful for internal spur
gears; the principle use of broaching in this context is for internal splines which
cannot easily be made by any other method. As with all broaching the method is
only economic for large quantities since setup costs are high.

The major application of broaching techniques to helical external gears is
that used by Gleasons in their G-TRAC machine. This machine ooperates by
increasing the effective radius of a hobbing cutter to infinity so that each tooth
of the cutter is travelling in a straight line instead of on a radius. This allows the
cutting action to extend over the whole facewidth of a gear instead of the
typical 0.75 mm feed per revolution of hobbing. The resulting process gives a
very high production rate, more suitable for U.S.A. production volumes than for
the relatively low European volumes and so, despite a high initial cost, is very
competitive.

Broaching gives high accuracy and good surface finish but like all cutting
processes is limited to "soft" materials which must be subsequently case-
hardened or heat treated, giving distortion.

V. SHAVING AND ROLLING

Shaving and rolling are both used as finishing processes for gears in the "soft"
state. The objective is to improve surface finish and profile by mating the
roughed-out gear with a "cutter" which will improve form.

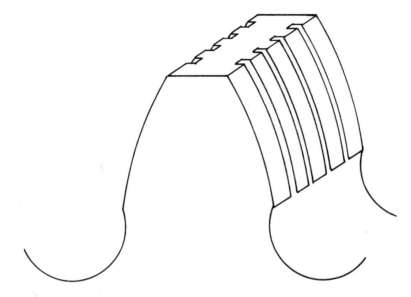

Figure 5.3.   Sketch of the shaving cutter tooth.

A shaving cutter looks like a gear which has extra clearance at the root (for swarf and coolant removal) and whose tooth flanks have been grooved to give cutting edges as indicated in Figure 5.3. It is run in mesh with the rough gear with crossed axes so that there is in theory point contact with a relative velocity along the teeth giving scraping action. The shaving cutter teeth are relatively flexible in bending and so will only operate effectively when they are in double contact between two gear teeth. The gear and cutter operate at high rotational speeds with traversing of the workface and about 100 micron (4 mil) of material is removed. Cycle times can be less than half a minute and the machines are not expensive but cutters are delicate and difficult to manufacture. It is easy to make adjustments of profile at the shaving stage and crowning can be applied. Shaving can be carried out near a shoulder by using a cutter which is plunged in to depth without axial movement; this method is fast but requires more complex cutter design.

Finish rolling, unlike shaving, does not remove metal but redistributes it by imposing sufficient local contact stresses to give plastic deformation. Figure 5.4 indicates the exaggerated profile shape before and after rolling; the surface is depressed about 50 microns and the surplus metal flows to regions at the root and tip, occasionally giving a burr at the tip if settings are not correct. Surface finish is good and accuracy is good with a long tool life but the tools are expensive and a rigid machine is required. There is debate as to whether the severe cold working of the surface gives a better or worse gear life. The principle disadvantage of the process is that the roughing, usually hobbing, process must be held to

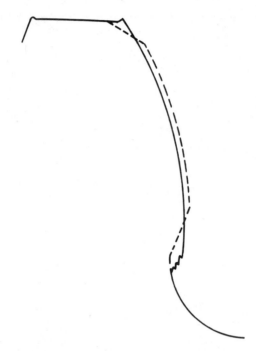

Figure 5.4.   Finish rolling. The dashed line shows the profile before rolling and the full line shows the shape after plastic deformation has occurred.

tight limits since rolling is less tolerant than shaving of errors in the rough gear. Economies at the finish rolling stage can easily be negated by extra costs at the hobbing stage: in general shaving is more economical for shorter and rolling for larger runs.

Rolling can also be used to form gears or splines from bar, bypassing the roughing process. Small teeth may be formed in ductile materials by this method which is popular for high production on small gears such as the gears used in epicyclios in automatic transmissions. For the larger gears used in manual trans- missions, with teeth 5 mm or more high, cutting methods are normally used. The production rates in rolling are high but the machines must be very rigid and the tools are expensive. In a single revolution of the working pair of tools, hundreds of teeth successively form the spaces; each tooth differs as the gear is formed and so each must be individually ground to shape. Splines are formed similarly but the formers may be racks instead of rotary tools.

Shaving and rolling both give high accuracy, usually after some "cut and try" at the initial design stage; however, with gears of very high helix angles, shaving gives errors due to deflections and shaving leaves existing pitch errors (1).

Subsequent heat treatment distortion must be allowed for at this stage but is often variable. Poor maintenance of shaving cutters tends in practice to reduce metal removal rates by factors of up to 4 (2) and lack of development of the infeed control rates on machines has also limited productivity. When high productivity is required the cutters must run true to within 5 microns and the shaving machines should be designed to be extremely rigid but to "break" and deflect easily when a critical force level is exceeded.

VI. GRINDING

Grinding is extremely important because it is the main way hardened gears are machined. When high accuracy is required it is not sufficient to pre-correct for heat treatment distortion and grinding is then necessary.

The simplest approach to grinding is form grinding, often termed the Orcutt method. The wheel profile is dressed accurately to shape using single point diamonds which are controlled by templates cut to the exact shape required; 6:1 scaling with a pantograph is often used. The profiled wheel is then reciprocated axially along the gear which rotates to allow for helix angle effects; when one tooth shape has been finished, involving typically 100 micron (4 mil) metal removal the gear is indexed to the next tooth space. This method is fairly slow but gives high accuracy consistently. Setting up is lengthy because different dressing templates are needed if module, number of teeth, helix angle, or profile correction are changed.

The fastest grinding method uses the same principle as hobbing but replaces a gashed and relieved worm by a grinding wheel which is a rack in section. Since high surface speeds are needed the wheel diameter is increased so that wheels of 0.5 m diameter can run at over 2000 rpm to give the necessary 1000 m/min (3000 ft/min). Only single start worms are cut on the wheel but gear rotation speeds are high, 100 rpm typically, so it is difficult to design the drive system to give accuracy and rigidity. Accuracy of the process is reasonably high although there is a tendency for wheel and workpiece to deflect variably during grinding so the wheel form may require compensation for machine deflection effects. Generation of a worm shape on the grinding wheel is a slow process since a dressing diamond or roller must not only form the rack profile but has to move axially as the wheel rotates. Once the wheel has been trued, gears can be ground rapidly until redressing is required. This is the most popular method for high production rates with small gears and is usually called the Reishauer method.

Large gears are usually generated by the Maag method which is similar to planing in its approach but uses cup grinding wheels of large diameter to form the flanks of the theoretical mating rack. Figure 5.5 indicates the orientation of the wheels relative to the roughed gear. Gears of very large diameter cannot easily be moved so the gear is essentially stationary while the grinding wheel

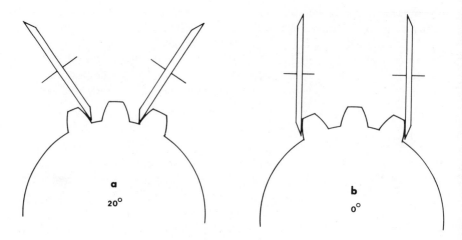

Figure 5.5. Grinding large gears. Cup wheels are used to give a flat surface in the position of the theoretical mating rack in approach (a) the 20° method. In (b) the 0° method the surfaces are parallel. The linear movement of the wheels relative to the work corresponds to movement at the base circle radius.

carriage reciprocates in the direction of the helix. The wheel is only in contact over a small part of the facewidth in helical gears so only a part of two flanks is being ground at any one time. The process is very slow but this is not important when only a few gears of this size are made in a year. As with form grinding, after grinding a pair of flanks the gear is indexed to the next pair.

A similar method used for medium size gears has stationary wheels, (which are effectively mating rack flanks) while the rough gear is traversed under the wheels. Corresponding rotational movement of the gear is controlled by steel bands unwrapping from a cylinder of pitch circle diameter so that the motion of gear relative to "rack" is correct.

Another method, the Niles approach, uses a wheel which is formed to give the "theoretical mating rack" instead of using two cup wheels as in the Maag method. This approach is best suited to medium precision work on smaller gears and is intermediate in speed between the Reishauer and Maag methods.

All grinding processes are slow and costly compared with cutting processes and so are only used when accuracy is essential. A rough rule of thumb is that grinding will increase gear cutting costs by a factor of 10 but the cost of the teeth is often only a small part of the total cost of a gearbox. The accuracies attainable are surprisingly not very dependent on size of gear; whether a gear is 5 m or 50 mm diameter the pitch involute and helix accuracies attainable are of the order of 5 microns (0.2 mils) or better and more dependent on the skill and patience of the operator and inspectors than on any other factors.

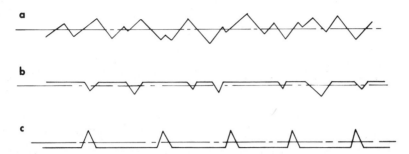

Figure 5.6. Surface finishes. (a) is a typical finish after grinding, (b) is a good
bearing surface suitable for gears or cylinder bores while (c) is the surface finish
required for abrasion.

One problem with grinding lies on the type of surface produced since
although great accuracy is obtained the surface finish is as shown diagrammatically in Figure 5.6(a) whereas the ideal surface for highly loaded contacts
is preferably as in Figure 5.6(b). Quoting surface roughness is no help since
surface roughness figures do not differentiate between a surface like Figure
5.6(b) and Figure 5.6(c); Figure 5.6(b) is an ideal bearing surface whereas Figure
5.6(c) is a very good file or abrasive surface. It is usually advisable to have some
form of polishing or burnishing operation after grinding to remove any sharp
tops of peaks though running in of gears may perform this operation.

It is often assumed that grinding will remove all errors generated at the
roughing stage. Unfortunately, grinding machines are relatively flexible and so
the grinding wheel has a tendency to follow previous errors. The errors will thus
be reduced but not completely eliminated unless very many cuts are used; whenever a grinding process is giving inconsistent results it is advisable to check the
accuracies at the rough-cut stage. The only exception is the form grinding
process which will not follow involute errors though it will still allow helix and
pitch errors.

VI. LAPPING AND HONING

As a final process gears in the hard state may be lapped or honed, mainly to improve surface finish and remove odd burrs. In lapping, loose abrasive is introduced into the mesh as a pair of gears are run together under light load.

Gears may be run either as pairs or individually with a master which is
often cast iron; if gears are lapped as a pair they should subsequently be kept as
a pair. It is not advisable to attempt to use lapping for profile correction as this
is liable to do more harm than good and it is unlikely that lapping will be able to
help much on very heavily loaded gears since contact is over a very small area

when lapping is occurring but is over nearly the full tooth flank under load. Heavy lapping of a pair of highly loaded gears would destroy the profile corrections and give high vibration levels under load.

Honing is a process akin to shaving but uses a bonded abrasive mating gear with coolant. More metal is removed than with lapping so that profile can be corrected as well as surface finish improved but helix and pitch are not usually much improved. Maintaining a sufficiently accurate form on an abrasive "gear" is not easy.

REFERENCES

1. Murphy N., Diagnosing unsuspected errors in shaved gears, Machinery and Production Engineering, 23, Paris 1975, pp. 384–386.
2. Welbourn D. B. and Smith J. D., Aspects of Shaping, Hobbing and Shaving machines for medium sized gears. Inst. Mech. Eng., Gearing in 1970 Conference, September 1970, pp. 1–9.

Chapter 6

GEAR MEASUREMENT

I. PROFILE, HELIX, AND PITCH

Figure 6.1 shows a sketch of a tooth flank on a gear; we normally measure profile and helix as indicated and measure pitch from the pitch point P at the middle of the face to the corresponding point(s) on the other flank(s).

Profile is measured in the transverse, not the normal, plane and the most straightforward method is to traverse a displacement transducer on an exact involute path relative to the gear so that any deflection of the transducer probe measures variation from the involute. An involute can be generated by rolling a straight edge, without slip, round the circumference of a disc which is exactly the base circle diameter; a point on the straight edge will then travel on an involute relative to the base circle. This method gives a basic accurate method of profile measurement but requires manufacture of a new base disc for each new design. Alternative methods use a master involute cam and an optical method of setting base circle radius (1).

Larger gears present problems in measurement since standard involute measuring machines can only take up to 1.6 m gears. Various rather clumsy methods have been used, either based on generating an arc or radius equal to the curvature of the profile at the pitch point or on measuring the tooth profile using a 3 coordinate system and calculating the correct involute profile. The first approach requires the use of linkages which are either located from the axis of rotation (2) or, like the profilometer, sit on the gear teeth (1) and since the radius required may be a metre, the instrument is large and not easily used. The 3-D method is either located relative to the axis of rotation as in the Maag approach or is a small unit mounted on the gear teeth as with the David Brown (3) or Vinco (1) design. The problem with the 3-D approach is not the calculations for conversion from 3-D to involute errors since although these are tedious they can be computed, but the cumulative inaccuracies due to location of the axes and the combined errors of measurement in 3 axes since all measurements influence the result. An alternative simpler and cheaper approach by the author to allow profile measurement on board ship uses the heresy of measurement of profile in

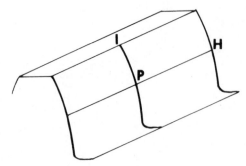

Figure 6.1. Sketch of tooth flank. P is the pitch point, PH is the helix measurement direction and PI is the profile direction.

the normal, not the transverse plane, and works by measuring deviation of the
tooth profile from the flank of the "theoretical mating rack" (4). As only one
direction of movement needs measurement accurately over a very short stroke of
100 microns the overall accuracy is high, direct readout is easily obtained and
the measuring head weighs only 1 kg; as speed of measurement is high ($<$ 5 secs)
it is possible to measure profile on a gear while it is being hobbed. The disadvantage of this approach is that a given instrument will only measure one normal
pressure angle; if 20° is normally used this does not matter and the very portable
nature of the instrument makes it convenient for monitoring damage or wear on
profiles in service.

Helix measurement presents fewer problems though it is necessary to
know where the axis of rotation is. Machines may either work by relating axial
movement to rotational movement, measuring both individually or by generating
a helix using a sine bar with linkage (bands or friction) to a base circle. It is
common to have profile and helix checking on the same machine since both require a linear movement which is equal or proportional to distance round a base
disc circumference. Scoles and Kirk (1) describe some of the instruments used.
Alternatively the 3-D rectangular coordinate approach may be used; the best
method with a 3-D machine however is not to attempt to measure helix directly
but to make an accurate measurement of axial pitch, to give the lead of the gear
and work back to the helix angle.

Pitch measurement may be of adjacent pitch, short spans of several teeth,
or cumulative pitch when circumferential pitch is required but occasionally base
pitch is measured between adjacent teeth. Measurement of base pitch, though
relatively easy and reliable is little used because variations in profile can give
deceptive results; consistency of base pitch round a gear is however a good measure of the consistency of load sharing between teeth in contact. Adjacent pitch
is simplest measured by a comparative technique using two probes, one fixed and
positioned in contact at the flank center on one tooth and the other spring

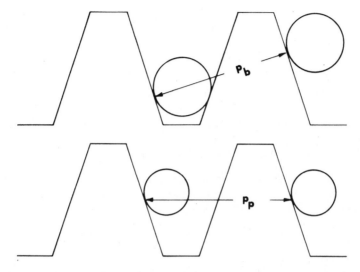

Figure 6.2.   Pitch measurement. p_b is the base pitch and is the shortest distance between successive flanks while p_p is the circumferential pitch and should be measured between points which are at the same radius from the axis of rotation.

loaded into contact with the center of the next flank. Figure 6.2 shows the ball positions for base and circumferential pitch checking. Absolute values of pitch are not usually required so it is sufficient to record variations in adjacent pitch and these may be summed to give cumulative pitch if required. Accuracy of this approach is not high since any measurement is subject to an uncertainty of the order of ±1 micron at best and this is often comparable with the gear errors. There may also be problems associated with mounting of the probe cradle if the top of the teeth or the teeth flanks are used as references since they may not be running true to the axis of rotation. Scoles and Kirk (1) describe several of the variants on this type of approach using two probes.

    If high accuracy is required for precision gearing it is preferable to use an optical method. Index heads or theodolites can easily be obtained with accuracies of less than 1 second of arc corresponding to 1 micron at 400 mm (16 inches) diameter. The problem then is to obtain an accurate stop to locate the pitch point exactly. Use of a mechanical stop tends to be unreliable since airborne dust may be 5 microns diameter and poor surface finish on gear can give 0.5 micron variation; it is advisable to use a noncontact pressure air probe for the reference as this is self-cleaning and insensitive to dirt and the restriction action of the nozzle averages the surface effects over a circle of 0.5 mm diameter. Re-traction of the probe for indexing to the next tooth can be done on a pre-loaded ball or roller sliding but is best on a linear air bearing for consistency.

"Home-made" equipment of this type can achieve 0.25 micron positioning accuracy with care. Large gears present problems since 1 second accuracy is no longer sufficient and it is then necessary to use two air pressure probes measuring adjacent pitch, one acting as a reference and the other calibrated to read variations in pitch (5).

3-D coordinate machines are not ideal for measuring pitch although they can be used for this purpose but it is more straightforward if a rotary axis of sufficient accuracy is available.

II. PROBLEMS WITH BASIC MEASUREMENTS

Profile, helix and pitch may be termed the "basic" measurement that we carry out on gears because we define the desired shapes of gears in terms of these three parameters.

Although these "basics" define our requirements, there are problems if we measure gears by these methods due to three main causes. The first one is an economics problem because the three measurements are relatively slow and require skilled labor; pitch measurement, particularly with shaped gears but, to some extent, with ground gears, must measure all teeth to be worthwhile and involute and helix checks should be taken on at least 4 teeth round the gear. We thus have the ridiculous situation that it may take only 5 minutes of machine time to manufacture a gear but it would take nearer 5 hours to check the "basics" of the gear fully.

The second problem arises from the arbitrary choice that we usually make in measuring profile and helix on lines which run through the pitch point. There could be large errors in the corners of the gear flank which would pass unnoticed unless extra profiles were taken near the ends of the gear, involving more time and effort.

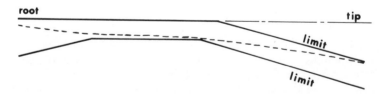

Figure 6.3. Profile limits. The full lines represent typical manufacturing limits. The dashed line is a profile which although within the specified limits would probably give high vibration levels due to insufficient tip relief.

Thirdly, the results obtained are still open to debate since it is quite possible for a gear profile or helix to be within specification and yet unsatisfactory as shown in Figure 6.3 or outside specification and satisfactory. In theory any criterion that a person uses to accept or reject a particular gear can be turned into a logical set of test parameters which a computer could use to carry out the same selection process but the state of the art at present is such that this, though academically feasible, does not appear to be economic.

III. ROLL CHECKING AND BEDDING

The need for a very fast, simple method for checking gears in mass production led to the use of roll (Parkson) checking, better called double flank checking. The gear under test is spring loaded into tight (double flank) mesh with a master gear or with its mating gear and rotated slowly while a dial gauge or transducer measures the variation in center distance between the gears. Tolerances are specified for the total movement during one revolution and the movement from tooth to tooth. Costs are low and testing takes only seconds so it is very popular as a quick check; it can pick out an eccentric gear or a poor profile which gives a once per tooth vibration or a "dropped tooth" pitch error.

Despite its popularity as a quick production check it is little used for development work due to its inherent defect of giving information which is the average of the information from the two flanks. Figure 6.4 shows diagrammatically an error which has occurred due to poor pitch accuracy on a gear manufacturing machine; this type of error will not be detected by a roll check since the effects on the two flanks cancel. Conversely a gear may be rejected due to a poor flank which is not used when installed. This basic problem of the averaging of information from two flanks cannot be overcome and is further complicated because for half the time (if the contact ratio is 1.5) there are theoretically 4 flanks in contact; in practice only two flanks are ever in contact but it is not possible to determine which two.

Bedding, though sometimes hardly considered as a "measurement" technique is an extremely important part of gear testing and must be carried out on any highly loaded gears. The pair of gears, correctly mounted and aligned, are run together after one gear has been thinly coated with toolmakers blue; the area of blue removed from one gear or marked onto the clean gear is a measure of the amount of contact that has occurred under no-load, i.e., inspection conditions. Bedding checks may be carried out under load by using lacquer or by lightly copper plating the tooth flanks and inspecting after running for a short time.

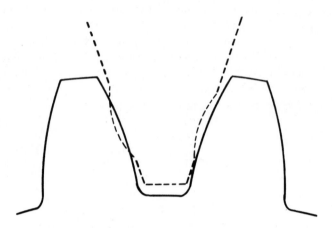

Figure 6.4.   Roll check error. The pressence of positive metal on one flank and negative on the other due to pitch errors will not show up since the effects cancel with double flank checking.

Bedding checks are very desirable because they are really the only way in which it is possible to confirm that gears are in full contact. The objective of the check, rarely stated, is to ensure that at all times in the mesh there is the maximum possible length of line of contact between the gear flanks. Any areas without contact show that increased loadings are being imposed on the rest of the line of contact. It is however important that the line of contact and, hence, the bedding pattern should not go right to the edge of the tooth either at tip or ends because the resulting stress concentrations at any unsupported edge would give rapid local failure and possible case separation. On heavily loaded gears, bedding checks which are carried out only in the no-load condition are liable to give deceptive results since the effect of deflections under load is to alter the contact pattern greatly but for gears which do not deflect much the no-load test is sufficient.

IV. TRANSMISSION ERROR

Noise and vibration problems in gearing are concerned less with the strength of gears than their smoothness of drive since it is speed variation and the consequent force variation that generates trouble. The corresponding research, development and production control problems are solely interested in smoothness and so it is logical to attempt to measure smoothness of drive directly. It is possible to estimate errors of drive by working back from the basic measurements but the results are far too inaccurate and very much work is involved. Bedding, though giving information on extent of contact, gives no information on smoothness.

The parameter measured is called Transmission Error (T.E.) and is defined as the difference between the position that the output shaft of a gear drive would be if the gearbox were perfect, without errors or deflections and the actual position of the output shaft. It may be expressed either as an angular displacement from the "correct" position or sometimes more conveniently as a linear displacement along a line of action, i.e., at base circle radius or as a linear displacement at pitch circle radius. Meshing equipment measures it in terms of angle and it is then convenient to have it as error along the line of action when considering profile modifications or at pitch circle radius when considering helix or pitch errors.

Aachen Technische Höchscuhle developed a seismic method of measuring torsional vibration at both ends of a gear drive. This equipment works effectively down to 1 Hz; it is suitable for measuring once-per-tooth errors but less suitable for once-per-revolution errors since inertia effects can come into play if unloaded gears of high inertia rotate at 60 rpm.

The most common method for T.E. measurement is to use radial optical gratings of very high accuracy attached to input and output. The input shaft is driven at a servo-controlled steady speed and the variation of light intensity through grating and indexes at once-per-line frequency gives variation of current through photodiodes to generate a regular sine wave at once per line frequency. Amplification and trigger circuits then give a pulse train whose frequency is multiplied and divided electronically to give the pulse train frequency expected at the output shaft. Measurement of the relative phase between processed input pulses and actual output shaft pulses gives the transmission error since variation of phase corresponds to variation of difference between expected and actual output shaft position. The equipment, developed from NEL work, is made by Gleason-Goulder and is accurate to a second of arc, far better than is normally needed. It runs at very low speeds, less than 10 revolutions per minute so there is no danger of dynamic effects distorting the results. Single pairs of gears may be tested or complete gearboxes may be tested; angled drives such as bevels or worms and wheels present no difficulty.

Transmission Error measurement using gratings came into existence primarily for the requirements of researchers on noise and vibration but it is now being used increasingly because it is an extremely fast method of checking and, hence, cheap. A pair of gears can be T.E. checked within a minute on both flanks so that 100% checking of gears can be achieved easily as the test is as fast as double flank roll checking, though far more accurate. In the future it is likely that the majority of parallel shaft gears will be checked this way with the "basic" methods only used for detail backup when setting up or when something is wrong. For bevels, hypoids, etc. it has virtually no competition. A major economic advantage of checking gears in pairs is that much less tight control on tolerances is required: 20% of a batch of gears may be individually outside a ±2 micron tolerance but then typically only 2% of pairings would be outside a ±4 micron tolerance.

V. USE OF T.E.

Transmission Error is of very direct use in noise or vibration investigations since it is the prime source of vibration in the system. For general use however it is necessary to decide exactly what information T.E. does and equally important, does not give. It is theoretically possible for gears to have a low T.E. yet to have such poor contact due to shaft misalignment that they have little strength. The best combination of tests is to determine T.E. to check smoothness of drive and then to check bedding to ensure full contact which infers satisfactory strength.

A T.E. trace is shown in Figure 6.5; it is similar to the trace obtained from double flank checking in that it results from variations in pitch, profile, and helix. In general there will be a fairly regular once-per-tooth pattern, particularly with spur gears, superimposed on larger waves due to once-per-revolution effects and sudden drops or rises due to adjacent pitch errors. Reversal of drive direction gives a trace from the opposite flanks which may well be very different in character.

Some pitch information can usually be deduced from the trace without complication if there are simple once per revolution waves which immediately suggest an eccentrically mounted gear. For any reasonably precise work, refer-

Figure 6.5.   Typical T.E. trace. A fairly regular waveform at once-per-tooth is superimposed on a once-per-revolution. There may also be high frequency vibration due to surface roughness effects.

ence shoulders should be provided on the gear and if the traces suggest eccentricity while the gear is running true then the problem often lies at an early stage of machining which produced eccentricity which was subsequently apparently shaved out or where inadequate grinding has followed previous errors. As any pitch errors on either gear will repeat at once per revolution of that gear it is easy to identify repetitive errors by their spacing and allocate them to one gear or the other. Slow rotation of the gears will locate the position of the defect on the gear and a local visual inspection may often discover a damage, burr, or impacted debris. A slightly more sophisticated, though not necessarily more effective, approach is to use a time averaging technique to separate out odd repetitive errors on the two gears.

A gear drive with a 1:1 ratio is often used in bevel gears and since errors on both gears occur at the same interval it is necessary to move one gear relative to the other to apportion pitch errors between the gears; traces are usually taken with 0°, 90°, 180°, and 270° relative movement of one gear and this allows separation of those effects which do not move from those which move relative to a once-per-revolution marker on the "fixed" gear.

Pitch information on a gear may be measured exactly by arranging to sample the T.E. signal each time contact occurs at the pitch point. Averaging the result obtained over several revolutions of a gear or ideally over a complete mesh cycle or pair of gears will give pitch figures which are independent of the accuracy of the mating or master gear. This allows measurement of gear pitch to an accuracy of much less than a micron even when meshed with a master gear which may have 5 micron errors provided the gears do not have the same number of teeth. If gears have the same number of teeth, averaging is much more complicated since each tooth in turn must be meshed with every tooth in the other gear.

Once-per-tooth errors occur in the trace almost invariably and are less easy to identify. Occasionally a tooth frequency error will appear as in Figure 6.6 with amplitude modulated at once per revolution frequency of either shaft (or both); this is easily identified as a gear which has been mounted or cut drunkenly on its shaft so that the effective helix angle varies. In general however if there is a regular once per tooth error it is not possible to specify which gear is

Figure 6.6.   T.E. with modulation. The error at once-per-tooth varies in amplitude during a revolution.

incorrect. Spur gears are designed to give tooth frequency error at no-load as shown in Figure 6.5 but, if the observed error is not correct, both gears must be individually checked by profile checking or against a master. Heavily corrected spur gears for high loads will normally have about 40% of their profile not in contact at low loads and so profile checks are normally needed to augment T.E. checks if center distance cannot be altered. When center distance can be extended, the upper part of the profile can be checked by setting the center distance to the point where the contact ratio is unity.

A correctly designed helical gear pair with a working facewidth of an exact number of axial pitches will in theory give zero T.E. if either the helices are correct or the profiles are involutes. There is thus a theoretical possibility of a pair of gears with involute gear profile meshing smoothly even though contact is only occurring at one end of the teeth or of incorrect profiles being hidden by a perfect helix match. The first is most unlikely to occur since gears are normally made with heavily corrected profile but incorrect profiles may well be averaged by helix effects since the main point of using helical gears is to obtain this averaging effect over the whole load range. This possibility can be checked quickly in inspection either by checking against a narrow master gear, say 6 mm wide, or by displacing one gear of a pair axially so that there is a small overlap instead of full facewidth contact; an implicit assumption is then made that the profile is consistent along the length of the tooth.

If a pair of helical gears has accurate helix matching but poor profiles, little will show on no-load T.E. and if the gears are mounted in their gearcase then it is not possible to give axial displacement. The alternatives for a thorough check are either to measure profile on both gears, a rather slow and expensive process, or to mount the gearbox in a test rig under full torque and measure transmission error and bedding; this method, though expensive to set up, is fast and gives the effect of shaft and gearcase as well as of manufacturing accuracy.

VI. INSPECTION CRITERIA

The "basic" measurements have been used for so long that a set of empirical rules have been evolved to classify gear accuracy. American, British, and German specifications quote permissible levels of profile, helix, and pitch errors for specified classes of gears and may specify bedding patterns. From the designer's point of view, particularly with noise in mind, it is important to remember that accuracy classes are completely arbitrary and may be very misleading. A particular gearbox installation may be very noise sensitive to tooth frequency and insensitive to eccentricity or vice versa. In one case a rear suspension happened to be very sensitive to twice tooth frequency and to once per revolution but not to tooth frequency. Increase of accuracy does not always improve performance of a machine and it is a well known trick to introduce pitch errors in some drives to reduce irritating whines.

Inevitably, specification of accuracy is a compromise in which cost plays an important part so it is worthwhile for a designer to specify the parameters individually instead of using a standard system. At very low speeds and high loads, for example, pitch and involute on a spur drive are often not important and can be several grades less accurate than helix. Conversely, for some drives in a gear cutting machine involute and helix are unimportant but cumulative pitch must be controlled tightly. A fundamental problem occurs with all accuracy specification as to whether full interchangeability is required and, if so, whether statistics may be used. As a simple example, an average loading of 140 N/mm (800 lb.f/in.) facewidth requires that the helices agree within ±10 microns to avoid 100% overload at one end of the teeth. Even if gearcase and shaft effects are neglected, 100% interchangeability could only be achieved by keeping both helices within ±5 microns, a very difficult specification for manufacturing. It is probably more economic to allow 10% of the individual gears to be outside specification and rely on bedding checks to identify the 3.5% of pairs of gears which will be out of tolerance.

When inspectors encounter T.E. measurements for the first time on helical gears they will usually ask what T.E. limits correspond to their familiar "basic" limits. There is no simple answer for profile and helix due to their interaction in T.E., although pitch tolerances, whether adjacent or cumulative, correspond more directly. In general, it is up to the development team to specify that the peak-to-peak error at once-per-tooth on the T.E. shall not exceed, say, 5 microns based on knowledge of the sensitivity of the overall design to excitation. Alternatively, it is better to specify the frequency components of the error so that limits are put on once-per-revolution, once-per-tooth, twice-per-tooth etc.; it is relatively easy to sample the output from the T.E. into a Fourier analyzer which will give the components immediately (6). In particular cases such as printing machines where register accuracy is important it is the peak-to-peak error which matters.

Spur gears do not show the interaction between profile and helix; in general, T.E. tests will give no information about helix accuracy. The relationship between T.E. and profile is discussed in a paper by Munro (4); from required profiles it is straightforward to predict what T.E. should be measured and so it is possible to check rapidly whether the "single tooth contact only" part of the profile is correct. When discussing measurements it is worthwhile confirming that the same terminology is being used by all concerned since "error" may mean either an inaccuracy in manufacturing or a perfectly legitimate correction to tooth form; an odd sounding case occurs when a person talks about the "error in transmission error" which means the variation from the (correct) transmission error expected from a pair of spur gears. A highly loaded gear may have an involute correction (tip relief) of 25 micron (1 mil) and when run with a similar gear would give some 8 to 10 micron T.E. at once-per-tooth frequency.

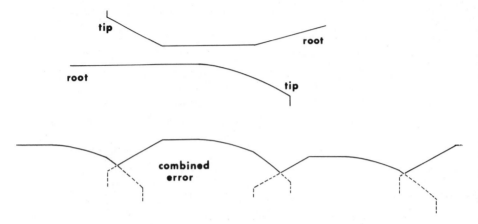

Figure 6.7.   Prediction of T.E. from profiles with spur gears. The meshing error for one pair of teeth is obtained by adding the profile errors at each point of the mesh relative to the pitch points. The relative heights of the pitch points for neighboring pairs of teeth are deduced from the pitch errors.

Figure 6.7 shows diagrammatically how the T.E. can be predicted with two tip relieved gears. The dotted sections correspond to those parts of the profile which do not come into contact at light loads.

REFERENCES

1. Scoles C. A. and Kirk R., Gear Metrology, MacDonald, London, 1969.
2. Bouillon G., Tordion G., and Tremblay G. Un nouvel instrument pour mesurer l'erreur de profil d'engrenages de grand diamètre. Trans. C.S.M.E., Vol. 1, No. 2, June 1972, pp. 97–100.
3. Watson H. J., Modern Gear Production, Pergammon, Oxford 1970.
4. Smith J.D., Involute measurement on large gears. Inst. Mech. Eng., 4th World Congress, Machine and Mechanisms, Newcastle, September 1975, pp. 215–218.
5. Sharpe J. E. and Smith J. D., Accurate pitch checking, J.M.E.S. Vol. 13, No. 2, 1971, pp. 116–118.
6. Thomson A. M., Fourier Analysis of Gear Errors, NEFLEX 80. N.E.L. Paper 3.5. (or from Gleason-Goulder)
7. Munro R. G., Effect of geometrical errors on the transmission of motion between gears. Proc. Inst. Mech. Eng. V. 184, 1969–1970. Part 3.0. pp. 79–84 (or from Gleason-Goulder).

Chapter 7

ELASTIC DEFLECTIONS

I. TOOTH DEFLECTIONS AND STIFFNESSES

The major step forward in the last 20 years in gear design has been a better understanding of the effects of elastic deformations on gear performance. Classic work by Harris (1) and Gregory, Harris, and Munro (2) established the existence of deflections and predicted the effects of elasticity on the vibrations of a drive system. This was followed by much work in Germany, summarized in a paper by Niemann and Baethge (3).

A pair of teeth in contact will deflect elastically due to local Hertzian contact deflections and due to the main stress pattern in the teeth. The local contact deflections are small compared with the tooth bending and shear deflections and are neglected; this appears arbitrary but arises from the small size of the contact patch which means that high stresses and the associated high strains do not penetrate far into the surface and so are unlikely to cause deflections of more than 5 microns. In contrast the stress pattern in a large tooth can give 50 microns deflection. Surprisingly the bending stiffness of a tooth is independent of its size and varies only with shape and with position of the force on the tooth flank. The stiffness of a pair of teeth remains reasonably constant because, as the contact position varies, the increased stiffness of one tooth compensates for the decrease in the other tooth. Steel gears of standard 20° pressure angle and shape were measured (2) as having meshing stiffness of 2×10^6 lb.f/ inch/inch facewidth and this figure, confirmed by finite element analysis, is the basis of design.

Taking a typical small gear of facewidth 15 mm (0.6 inches) subject to 2000 N (500 lb. force) load the standard stiffness of 1.4×10^{10} N/m/m facewidth (2×10^6 lb.f/inch/inch) gives a deflection of 10 microns (0.4 mil). If one contact is carrying all the load it will deflect 10 microns and the next pair of teeth will come into contact with an initial 10 microns interference unless allowance is made. This initial interference occurs at a tooth tip and would cause high local stresses with severe local damage. To avoid interference the teeth depart from a pure involute so that the initial "contact" at the tip has clearance then comes into contact smoothly away from the tip.

The amount of tip relief needed is often based on experience but may be estimated simply. It is the sum of the tooth deflection, the manufacturing base pitch error, and the variation of profile from tooth to tooth; this sum gives the worst case condition. Typically, for the gear mentioned above, the elastic deflection of 10 microns would be added to a maximum manufacturing adjacent pitch error of 8 microns *for each gear* to give 26 microns tip relief. The allowance for variation in profile is usually very small with hobbing or grinding, less than 2 microns, giving a total of 28 microns. Care however must be taken in assessing the elastic deflection since if the gears are misaligned, i.e., there are effective helix errors, the loading can easily be 50% above nominal at one end of the tooth giving an additional 5 microns in this case. It is customary to give tip relief from the tip down nearly to the pitch radius as indicated in Figure 7.1 and occasionally to give root relief. There is no inherent difference between giving tip relief and giving root relief but extra care must be taken with root relief as the geometry is more complex particularly with small numbers of teeth. In some extreme cases with low contact ratios root relief can destroy the effects of tip relief so it is preferable to give only tip relief.

The typical figures given above are interesting as they show the very large effect of adjacent pitch errors in both gears. If two successive contacts do not have a pitch error and the gear is lightly loaded there will apparently be no contact until the end of the tip relief is reached; this however neglects the geometry of the other contact which will be encountering its tip relief. Figure 6.7 of Chapter 6 showed how an idealized profile with tip relief would mesh with a similar profile at low loads and how the combined displacement curve could be deduced from the profiles. The effect of large tip relief, to allow for heavy loads and pitch inaccuracies, is to give vibration at low loads with spur gears. At high loads, the tooth deflections should be such that they cancel the tip relief effects exactly.

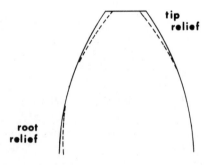

Figure 7.1.   Sketch of tip and root relief deviations of the profile from an involute.

The other variable in tip relief is the distance down the profile to the end of tip relief. This should be set by the contact ratio since the ideal requirement is for one contact to start relieving the load just as the other starts taking load; while only one pair of teeth are in contact there should be no tip relief since departure from the involute during the single contact condition only serves to generate vibration. Tip relief should not extend to the pitch radius, unless the contact ratio is 2, but typically for 1.5 contact ratio, should extend about a third of the way down the flank. Specification of "distance" down the profile is probably best in terms of roll angles since the angle turned through by the gear as a contact point moves along the pressure line gives the clearest mental picture of the movement of the system.

A further objective of tip relief is to give a smooth transfer of force from one tooth to the next; any abrupt change of position of a force tends to give high frequency vibration so the force changeover time should be as long as possible, consistent with not encroaching on the single contact part of the profile. There is, however, with helical gears a fundamental clash between the requirements for noise and strength since, although noise reduction prefers an extended tip relief to give the smoothest possible changeover, strength requirements want minimum extent of tip relief to keep the effective length of line of contact as high as possible to reduce stresses. This infers that different profiles and, hence, different cutters should be used when cutting spur and helical gears if strength is important.

II. TOOTH END EFFECTS

The objective of profile tip relief is to get a smooth force transfer with no stresses on the unsupported tip of the tooth and there is a corresponding requirement to achieve a smooth transfer of force at the ends of the teeth.

It would be dangerous to allow the line of contact on a gear face to extend right to the end of a gear since there would be high local stresses at the corner of the tooth, leading to case spalling or local fatigue damage so it is customary to give end relief to highly stressed gears so that the load distribution along the face is as sketched in Figure 7.2. The amount of relief can be estimated as the tooth deflection, allowing for any effects of helix error or misalignment. As with tip relief a linear decrease in end relief is reasonable but it is difficult to decide how far along the tooth should be relieved since end relief decreases effective tooth width.

On a spur gear, local stressing is the only consideration so the end relief can be very short in the form of a chamfer, though it is better to have a blended curve since there is otherwise a stress concentration where the chamfer ends. The ideal is a smooth transition similar to that used at the ends of rollers in anti-friction bearings.

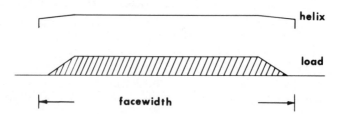

Figure 7.2. Sketch of end relief on spur gear tooth and corresponding load pattern.

Helical gears are rather more complex and require care in the selection of end relief. In Figure 7.3, viewing the pressure plane, a line of contact in the lower right hand corner is about to run off the corner of a tooth and disappear. The combination of tip relief and end relief on the gear will give a triangular load distribution as shown; the total force on the tooth corner will be low but only acts on a short section of tooth and the unloaded section of tooth will buttress the corner. The stress distribution is sufficiently complex that only finite element techniques can provide an answer. If no end relief is applied on a gear with a low helix angle there is some load reduction towards the end due to a reduced tooth stiffness, but with a high helix angle there is a load intensification effect. This will not increase root stresses but can increase Hertzian pitting stresses. Common sense suggests that end relief of about the normal module will be sufficient.

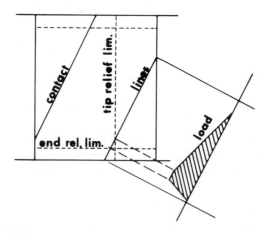

Figure 7.3. Load distribution with helical gears. In the view of the pressure plane one contact line is shown with load reduction at one end due to tip relief and at the other end due to end relief.

A clear distinction should be made between end relief and crowning which is sometimes used in cases of misalignment and extends along the whole face of the gear. Crowning should never be used for highly loaded gears (except as a small correction. See Section IV.) since the effect of large amounts of crowning (greater than 50 microns) is to give localized loading; it should only be used for gear tooth couplings where surface ratings are low.

In Chapter 3, Section IV it was noted that it was advisable to have the face-width as an exact number of axial pitches to keep the length of line of contact and, hence, the contact stiffness constant. End relief reduces the effective face-width by half the amount of end relief at each end so allowance should be made for this; under low loads the effective facewidth reduces but this does not matter since the deflections are low.

A further refinement in helical gear design proposed by Rouverol (4) is concerned with the problem of keeping the position of the resultant gear contact force constant as well as keeping the length of line of contact constant. The line of action of the resultant force in a gear mesh can be found by locating the center of pressure of the contact lines as shown in Figure 7.3. The center of pressure moves slightly axially as the mesh occurs and this can theoretically give rise to vibration. Cutting away part of the working flank of the gear as indicated in Figure 7.4 will give a constant position for the force but it is unlikely that this refinement will be needed in normal gears since effects due to force amplitude variation are so much larger than those due to position variation.

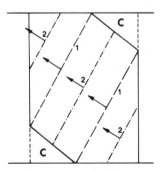

Figure 7.4. Silhouetting. View of pressure plane. The tooth corners corresponding to the corners C are removed to give constant length of line of contact as well as constant position of line of thrust. Lines 1, dashed, are the contact lines at one instant and lines 2, chain dotted, are the contact lines half a tooth later.

III. HELIX ACCURACY EFFECTS

On spur and helical gears the effect of misalignment of the mating flanks of a gear is to increase loadings at one end of the gear. It is interesting to take some typical values to see how large the effects are; a spur gear mesh with 9×10^4 N/m (500 lb./in.) loading will deflect 6 microns with one pair of teeth in contact. A corresponding misalignment of the mating flanks of 12 microns will give no contact load at one end and double loading at the other end so the system is extremely sensitive to alignment errors, whether due to gear helix angles or poor mounting or system deflection.

As misalignment can be contributed by both gears, each gear should be accurate within 6 microns if 100% overloads are to be avoided and this figure should cover the three possible sources of error. This accuracy cannot be achieved since axis alignment to such accuracy is rarely possible and thermal effects in gear cutting machines can easily produce this level of error. It is not usually appreciated by gear engineers how large an effect misalignment gives and it is probably misalignment which has been the major cause of the very large safety factors applied in gearing in the past. Turning the figure of 6 microns into an angle gives a tolerance of 40 seconds of arc on a 30 mm wide gear or 4 second of arc on a 300 mm (12 inch) wide gear. The extreme sensitivity of stressing to helix matching reinforces the importance of bedding checks which are the only reliable test; even bedding presents problems since the applied film is liable to be appreciably thicker than 12 microns.

When the misalignment of flanks is less than the tooth deflection as sketched in Figure 7.5(a) the amount of overload is easily estimated as it is half the total misalignment divided by the nominal tooth deflection. High misalignments give even greater overloads as the width of contact decreases as shown in Figure 7.5(b); the estimates are slightly more complex since, initially, the extent of contact is unknown. A typical example: 50 mm facewidth, spur gear, 3000 N total load with a total combined misalignment of 25 microns.

The nominal tooth deflection is 60×10^3 N/m divided by 1.4×10^{10} giving 4 microns so contact will be lost over much of the facewidth. Taking the extent of contact (as shown) as d meters and the maximum loading as L N/m then, $0.5 \ dL = 3000$ N and 25×10^{-3} is the slope which is also $L/1.4 \times 10^{10}$ divided by d, so that $L/d = 7 \times 10^6$. This gives d as 29.3 mm and the peak contact loading as 2.05×10^5 N/m, i.e., 205 N/mm which is over three times the nominal loading. The facewidth in this case could be halved without significantly increasing stresses.

Vibration levels in spur gears are usually little affected by helix errors and may be improved since helix errors can produce reduced contact length and, hence, reduced contact stiffness. The transmission error involved will not alter with alignment and, hence, T.E. testing will not give indication of trouble with helix.

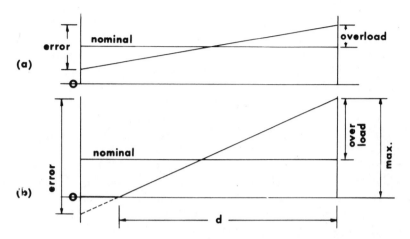

Figure 7.5. Helix error effects on loading: (a) is for a condition of error less than twice nominal deflection where contact is maintained across the tooth. (b) is for high errors where contact only occurs over a facewidth d. Loading is proportional to deflection except when contact is lost.

The situation in helical gears is complex since involute and helix interact in the transmission error. Good alignment of the flanks will give very low T.E. and, therefore, very low vibration levels regardless of the load or profile provided the facewidth is correct; this is the great advantage of helical gears. Poor alignment, as in the spur gear example above, has the effect of throwing the contact to one end of the gear so that the gear tends to behave rather like a spur gear and generate once-per-tooth effects because averaging across the facewidth is no longer occurring.

At light loads contact will occur only at one end of the gear and the transmission error will be exactly the same as a spur gear with once-per-tooth errors of, typically, about half the amount of tip relief present. As the load increases, the width of contact correspondingly increases so that there are some averaging effects across perhaps half the tooth width and the load per unit width on the end of the gear will reach the correct value for the particular involute profile corrections. This combination will give a relatively low T.E. and a correspondingly quiet drive although the length of line of contact will be varying appreciably; this condition might well occur at about quarter load. At full load the end of the gear will be heavily stressed so that the profile correction is unlikely to be sufficient and although some averaging will occur the vibration will be powerful. It will never be as powerful as the corresponding spur gear vibration in the tangential direction, although, due to the helix angle, it can generate axial vibration.

IV. GEAR BODY DEFLECTIONS AND DISTORTIONS

We have so far considered gears as teeth, elastic in bending, mounted on a rigid immovable body but in practice the body of the gear is not rigid and it is necessary to estimate and allow for any movement. There are three dominant causes which should be investigated in high performance designs; they are twist, whole body deflection, and bending. The deflections are calculated separately,   superposed, and equal and opposite corrections applied to the gear.

Twist of a gear is the simplest effect and occurs most with small diameter, high facewidth, highly loaded pinions which are driven from one end only. Standard shaft torsion formulae can be used and it is assumed that the applied load is evenly distributed; polar moment of inertia for the stiffness can be estimated using the root diameter of the teeth since the teeth themselves have little effect. As an example take a spur pinion of 12 mm module with 24 teeth and a facewidth of 0.4 m transmitting 60 kW at 30 rpm in a machine drive.

The transmitted torque T is $60 \times 10^3/2\pi 0.5$ which is 1.91×10^4 N m and this torque acts on a cylinder of effective diameter $[24 - (2 \times 1.15)] \times 12$ mm, i.e., 130 mm radius. The torque varies along the cylinder linearly with distance

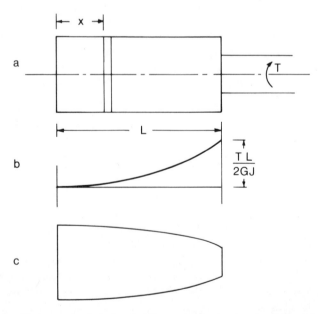

Figure 7.6.   Pinion twist (wind up) effects: (a) is the gear loaded with a torque T, (b) is the resulting torsional deflection along the gear, and (c) is the shape of a gear tooth corrected on both flanks to give constant loading across the facewidth.

as shown in Figure 7.6(a) and working from the "free" end the torque at distance x along is Tx/L so the angle of twist locally is $\delta x \times$ Tx/L divided by GJ, the shear modulus times the polar moment of inertia. Integrating from x = 0 to x gives the twist angle as $Tx^2/2GJL$, i.e., parabolic with a value of TL/2GJ at the end as sketched in Figure 7.6(b). In this case the value is $1.91 \times 10^4 \times 0.4/2 \times 81 \times 10^9 \times 0.5\pi .13^4$ which is 1.05×10^{-4} radians and at pitch radius of 144 mm this is 15 microns. The gear should thus be made such that the free end contacts first, there is a relief of 3.75 microns at the center and 15 microns at the shaft end to allow for windup effects. If the gear only drives in one direction the other flank of the gear is not important but if full power reverse drive is required a similar correction should be applied to the other flank giving a tooth shape as sketched in Figure 7.6(c) with the teeth at the free end of the gear 30 microns thicker than at the shaft end.

Whole body deflection occurs most dramatically when a gear is overhung on a shaft as shown in Figure 7.7. Even reducing the overhang to zero still leaves problems as the shaft bends between the bearings allowing tilt of the gear. When the gear is appreciably larger diameter than the shaft, bending within the gear can be ignored and shaft deflections dominate. Estimation of the tilt angle of the gear due to shaft and bearings is done by adding the effects of shaft and bearing deflections; the tilt due to the bend between the bearings is given by calculating the deflection due to the rear bearing force and using this to determine the gradient of the shaft through the front bearing. The expression obtained is:

$$\theta = \frac{Wb}{K}\frac{1}{a}\frac{1}{a} + \frac{W(a+b)}{Ka}\frac{1}{a} + \frac{Wb}{a}\frac{a^3}{3EI}\frac{1}{a} + \frac{Wb^2}{2EI}$$

K is the bearing support stiffness and EI is the local shaft bending stiffness. As an example take a gear with a total tooth load of 10,000 N (1 ton) overhung with center 75 mm from the front bearing, bearing spacing 300 mm, shaft diameter 80 mm between bearings, and 60 mm for the overhang; the bearing support stiffness is 10^9 N/m (5×10^6 lbf/in).

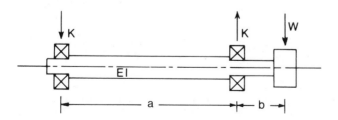

Figure 7.7. Gear body deflection due to shaft bending. Applied load is W, bearing stiffnesses K, and shaft bending stiffness EI (different for sections a and b).

$$\theta = \frac{.075}{10^4} + \frac{.375}{.3^2 \times 10^9} + \frac{.075 \times .3}{3 \times 210 \times 10^9 \times 0.25 \, \pi \, (.04)^4} + \frac{.075^2}{2 \times 210 \times 10^9 \times .25\pi \, (.03)^4},$$

i.e., $\theta = 8.3 \times 10^{-6} + 41.7 \times 10^{-6} + 178 \times 10^{-6} + 210 \times 10^{-6} = 4.38 \times 10^{-4}$.

It is interesting that despite the larger diameter of the shaft between the bearings the third term which is the contribution of the bend between the bearings is nearly as large as the tilt due to the overhung section.

The slope of 4.38×10^{-4} radians on a gear of facewidth 65 mm would give an error of 28 microns, far larger than any manufacturing errors; the change of helix angle involved is 1.5 minutes of arc. As with the twist correction, torque reversal would require correction of both flanks to give a gear tooth 56 microns wider at the outboard end. The terms due to the bearings are relatively small in this case though not always; unfortunately it is extremely difficult to get reliable information on bearing stiffness and as the support casing deflects as well, it is advisable to carry out some static deflection tests on the gearbox.

Figure 7.8 shows a gear which has been mounted unsymmetrically between bearings and will be tilted; in this case the tilt is in the opposite direction. Again superposition can be used to determine slope. All problems which do not involve redundant supports, i.e., more than two bearings, can be solved by superposition of six standard cases of cantilevers of length L, bending stiffnesses EI, loaded by a moment M, or a concentrated load W at the free end, or by a distributed load w per unit length.

Load	End slope	End deflexion
Moment M	$\dfrac{ML}{EI}$	$\dfrac{ML^2}{2EI}$
Concentrated W	$\dfrac{WL^2}{2EI}$	$\dfrac{WL^3}{3EI}$
Distributed w	$\dfrac{wL^3}{6EI}$	$\dfrac{wL^4}{8EI}$

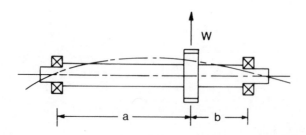

Figure 7.8. Gear body tilt between bearings.

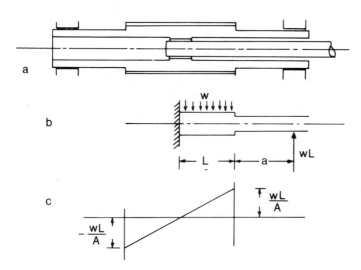

Figure 7.9.   Pinion bend and shear. (a) Pinion supported in bearings and driven centrally by a quill shaft so that torsion is small. For bending deflections, half the pinion is considered as sketched in (b), loaded as shown. (c) Shows the average shear stresses within the pinion body due to the force system.

In the case in Figure 7.8 the end deflexions relative to the tangent to the centerline at the gear are $W a^3 b/(a+b)3EI$ and $W b^3 a/(a+b)3EI$, so the slope is $W ab(a^2-b^2)/(a+b)^2 3EI$, i.e., $W ab(a-b)/(a+b)3EI$.

Bending within the body of a gear is usually only important with wide pinions which are heavily loaded. A typical case is sketched in Figure 7.9(a) where the pinion is supported symmetrically. As in the previous section it is assumed that the bearings act as simple supports and do not restrain bending. To estimate the deflexions within the gear it is simplest to consider the system of Figure 7.9(b) as if the gear were built-in at its center. It is then subject to a distributed load w, a shear force wL at the end, and a bending moment wLa at the end. The total deflection at the end of the gear is $WL^4/3EI$ plus $wLaL^2/2EI$ less $wL^4/8EI$.

Taking figures of 200 N/mm (1200 lb/in) loading, gear half length 150 mm and effective diameter 200 mm with a shaft length to bearing of 125 mm gives the gear distortion as $2.10^5 (.15)^4/3.210 \times 10^9 .25\pi(0.1)^4 + 2.10^5 (.15)^3 \times .125/2.210.10^9 \times .25\pi(0.1)^4 - 2.10^5 \times (.15)^4/8.210 \times 10^9 .25\pi(0.1)^4$ which is 2.04 + 2.56 − 0.77 = 3.88 microns. This is a very small deflexion and is usually ignored; it should be compared however with a mesh elastic deflection of 14 microns. The distortion is roughly parabolic in shape and unless corrected will give about 18% higher loadings at the ends of the teeth.

There are also elastic deflections within the body of a gear due to shear stresses. The shear stress at the gear center (Figure 7.9(c)) is zero by symmetry and increases linearly to the end of the gear. At distance x from the center the average stress is wx/A where A is the cross sectional area so the deflexion is the integral of wxdx/AG giving a parabolic distribution to 0.5 wL²/AG. Substituting for the gear above gives $0.5 \times 2.10^5 \times (.15)^2 / \pi (0.1)^2$ 81 × 10^9 = 0.9 micron, which is negligible.

From the examples given it can be seen that overhung gears are very susceptible to tilting quite large amounts and should never be used when loadings are high. Twist or wind up is usually larger than bending with pinions but can be very greatly reduced by using a drive to the center of the gear as shown in Figure 7.9.

Bending deflections are not usually significant with gears whose diameter is greater than their facewidth but particular designs may need very broad, small diameter gears. The most economical way to provide a rack drive with forces of the order of 30 tons is to use pinions whose diameter is about 200 mm and facewidth 400 mm so large corrections are required when loadings are of the order of 700 N/mm (2 tons/inch) facewidth.

These calculations of deflections and, hence, corrections are based on simple stress patterns and so will not give exact answers; they are sufficiently accurate for most purposes but care should be taken when particular aspects of design affect deflections. Figure 7.10 is a sketch of a typical pinion to wheel mesh where the wheel has a relatively narrow web. For this design it is theoretically possible to adjust the wheel rim flexibility to give exactly the right

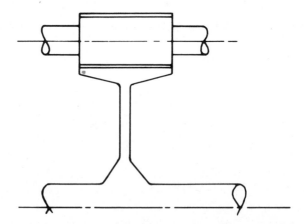

Figure 7.10.   Sketch of pinion and wheel. The wheel rim is elastic and will deflect radially due to the inward component of the tooth forces. This can be arranged to balance distortions in the pinion due to bend and twist to a first order.

elastic crowning on the wheel to match the crowning needed for pinion twist and pinion bending so that the only machining adjustment made is the change of helix necessary for the tilt effect due to twist in the pinion.

Corrections have in all cases been estimated for full load condition so that at part load for fully corrected gears there will be bad load distribution on the teeth since deflections are low. The loadings will not exceed the design load so that extended part load running will not affect the overall life of the gear though it may lead to selective pitting on the teeth at the points of maximum correction. The other "side effect" of corrections to the helix to allow for distortion is that at part load the meshing will be at one part of the face only so that transmission error will be higher than at full load as the averaging effect of the helix is reduced.

V. AXIAL FORCE EFFECTS

Single helical gears, unlike spur or double helicals, have an axial force which can cause movement and tilt. Movement is not important but tilt effects can be significant.

The worst case for this type of distortion is shown diagrammatically in Figure 7.11 where a gear is mounted symmetrically in a rigid case but the axial force component at pitch radius tilts the gear. The bending moment imposed at the shaft center is Wr and this produces end forces Wr/2a and a corresponding tilt of Wra/6EI radians. As an example take a loading of 170 N/mm (1000 lb./inch) facewidth on a gear 50 mm wide and 240 mm pitch diameter supported on a 50 mm diam shaft with a total span of 0.44 m between bearings. The pitch helix angle is $30°$. The axial force component is $170 \times 50 \times \tan 30° = 4907$ N so that the tilt is $4907 \times .12 \times .22/6 \times 210.10^9 \times 0.25\pi \times (.025)^4 = 3.35 \times 10^{-4}$ radians and the corresponding deflection is 16.8 microns across the gear face. This is a deflection in the radial direction so the corresponding tangential correction is $16.8 \tan \phi_t$, which is about 7 microns. Deflections of bearings could add to this deflection but it is most likely to occur with large gears on slender shafts. Correct positioning of a gear between bearings can often be arranged to make

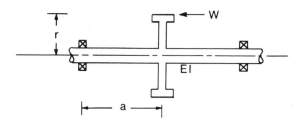

Figure 7.11. Axial force effects. The effect of the axial component W is to tilt the wheel giving misalignment of the mesh.

the tilt effects due to the axial force cancel out the tilt effects due to the tangential component of force.

VI. THERMAL EFFECTS

Thermal expansion may be ignored on small gears but with a typical coefficient of expansion of 10^{-5} per °C a gear of diameter 0.4 m will expand 3 microns per °C so that even 5°C gives 15 micron growth. Overall growth is unimportant but a 5°C differential between the center and the end faces of a gear of this size will be significant since 15 microns corresponds to a tooth loading of 210 N/mm facewidth (1200 lb/inch).

The conductivity of steel is low, 54 W/mK (30 Btu/sq. ft./hr./°F/ft.) so that low powers give quite large temperature differentials; in the particular case of a cube of steel 100 mm in size, the conduction between two opposite faces is 5.4 W per °C. A planet gear supported on a hydrodynamic bearing will be subjected to heating from the bearing as well as from the churning of oil in the tooth contact and total losses are typically 2% of the transmitted power so that a total heat generation of the order of 20 kW must be catered for in a 1000 kW gearbox. Roller bearings under optimum conditions of loading and lubrication, rarely achieved, have coefficients of friction of the order of 0.003 giving significant heat generation.

It is very difficult to predict temperature distribution in a gear with any accuracy since usually little is known about the generation of the heat and the effectiveness of the cooling so it is probably worthwhile instrumenting a prototype to check temperature distributions in large boxes.

In extreme cases when large transient temperature differentials are involved due to neighboring heat sources such as turbines the expansion involved can be sufficient to destroy backlash and give double flank contact with rapid failure. A gear of radius 200 mm requires 50°C differential to take up a backlash margin of 100 μm. This condition is unlikely to occur during running since air movement gives increased heat transfer but can occur during warmup procedures and has caused a failure in a gear tooth coupling.

REFERENCES

1. Harris S. L., Dynamic loads on the teeth of spur gears. Proc. Inst. Mech. Eng., Vol. 172, 1958, pp. 87–112.
2. Gregory R. W., Harris S. L. and Munro R. G., Dynamic behaviour of spur gears. Proc. Inst. Mech. Eng., Vol. 178, 1963–64, Part I, pp. 207–226.
3. Niemann G. and Baethge J., Drehwegfehler, Zahnfederhärte und Gerävsch bei Stirnrädern. V.D.I.-Z. Band 112, 1970, Nr. 4, pp. 205–214 and Nr. 8, pp. 495–499.
4. Rouverol W. S., Constant tooth-load gearing. World Congress on Gearing. Paris, June 1977, pp. 257–268. (Rolling Contact Gear Company, Berkeley, California.)

Chapter 8

NOISE GENERATION

I. CAUSES OF NOISE

There are fundamentally three possible ways in which noise can be generated from tooth contact forces, as mentioned in Chapter 1. If the force at the mesh varies in amplitude, direction or position there will be vibration.

If as in Figure 8.1(a), two discs, one with a corrugated surface, run together there is a vibration generated in the direction of the line joining the centers and noise is generated as when driving over a rough road surface. In Figure 8.1(b), the surface corrugations are on the teeth surfaces or due to errors between teeth so the vibration generation is in the direction of the line of thrust. The force between the gear teeth represented by the double-ended arrow acts on the gears which are in turn supported on elastic shafts and bearings in a gearbox which is itself a complex system which can vibrate to radiate sound directly or can transmit vibration through its supports to the remainder of the structure. Variation of force at the mesh excites vibration which it is convenient to split into internal responses (sometimes at resonances) which are of relatively large amplitude at the gears and these excite case responses which are usually smaller in amplitude.

Of the three possible causes we can eliminate variation in direction of force if involute gears are used since the common normal, that is the direction of the pressure force, does not change direction. There is a minor residual cause of change of direction of the force if there is friction which reverses direction at the pitch point: a coefficient of friction of 0.05 would give a direction change of six degrees in the force which in a typical gearbox would give the same vibration level as a 10 percent change in force amplitude. Since theory predicts a vibration due to change in friction direction, experiments have attempted to check whether this effect exists with contradictory results (1,2). The friction reversal can be eliminated by arranging that corrected gears are used so that an all addendum gear mates with an all dedendum gear; the disadvantage of such gears is that sliding velocities are much increased so the possibility of scuffing is increased. The reasons for the difficulties encountered in experimental checking of this possible effect are probably due to the very low level of the friction

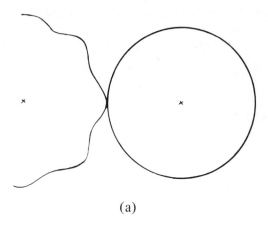

(a)

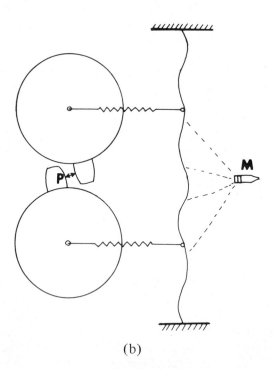

(b)

Figure 8.1. (a) Effect of surface undulations. (b) Noise generation. Relative displacements at P due to T.E. vibrate the gears and the resulting force variations transmit via the bearings to the case which when excited radiates noise to the observer.

forces compared with the force variations due to gear inaccuracies; an error of one micron will give greater vibration than possible friction reversal effects.

Friction reversal or pitch order impulse noise is unimportant in practice because effects cancel in helical gears; the cancellation is not complete, even in a well designed gear since there is a small residual torque but the effects may be ignored. Since pitch friction reversal could only be relevant for spur gears which are unlikely to be used for very high precision work, there is little point in putting effort into reducing low levels of spur gear friction reversal noise when it is so much easier and more effective to use helical gears.

Variation in the distances of the force from the axis of rotation will not occur since the force always acts in the pressure plane but the force can move axially along the gears a small distance giving the effect of a moment superposed on a steady force. The movement axially of the line of action is at its worst when the contact ratio is 1.5 and the facewidth is about an axial pitch so that the contact lines are as shown in Figures 8.2(a) and 8.2(b). For the worst case, the change in position is one sixth of the facewidth and so gives a worst variation of one sixth ($\pm 8\%$) in the force at bearings when the gear is supported at its ends. In most cases a transmission error of 1 micron will give a much larger force variation at the bearings so it is rare for this effect to be significant.

It can be eliminated by the technique of "silhouetting" as described in Chapter 7, Section II.

The other minor possible cause of noise comes from the transfer of load from tooth to tooth. This effect can be significant in very large gearboxes where the load transfer time becomes comparable with the time taken for stress waves to travel from one tooth to the next and is most likely to occur with a large thin

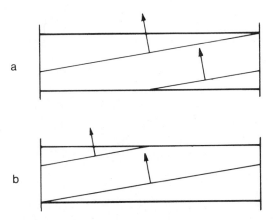

Figure 8.2. Variation of force position. As the contact line pattern changes from a to b each tooth the line of action of the resultant force moves axially, i.e., across the facewidth.

annulus where bending waves travel relatively slowly. Such problems are unlikely to be encountered.

An unusual form of gear noise can occur if there is a generous oil supply to the inlet side of the mesh with spur gears of high facewidth. It is then possible for oil to be trapped in the roots of the teeth and develop high pressures when it cannot get out fast enough at the ends of the teeth. The high pressures generated in the oil give a once per tooth excitation forcing the gears apart and the ejection of oil at high velocity at the ends of the teeth can give impacts of oil jet on the casing, again at once per tooth frequency. Properly designed helical gears are unlikely to suffer from this effect since even if oil trapping occurs there is an averaging effect along the helix which gives steady forces and it is relatively easy for the oil to escape when helix angles are high.

The effect of surface finish on gear noise is debatable. It is unlikely that surface finish will have a direct effect on T.E. and hence on vibration since the short distances between the peaks of surface roughnesses correspond to frequencies far above tooth meshing frequencies. The only effect observed would then be a "white noise" at frequencies outside the audible range. There will be an effect of roughness on local surface stresses, leading to pitting and there will be more likelihood of metal to metal contact through the oil film; it is difficult to see how this will increase noise without surface damage occurring.

We are left with Transmission Error as the dominating cause of gear noise and so can consider a gearbox as a vibrating system which is excited at the gear mesh by a T.E. which is a relative displacement between the gear teeth. The details of the T.E. are often not known at the design stage so comparison of designs is best done by estimating vibration levels that will be excited by a regular 1 micron error at the mesh, particularly at tooth frequency and its harmonics.

II. SIMPLE IDEALIZATION

Modelling of gearboxes is possible by reducing the system to a relatively small number of lumped masses connected elastically; this type of modelling is not suitable for large gearboxes but is satisfactory for most applications.

A typical machine tool gearbox will have solid gears supported on relatively long slender shafts in a gearcase which is much more rigid than the shafts. The corresponding model for two shafts is fairly simple and is shown in Figure 8.3; each gear has mass and moment of inertia and is supported on the lateral stiffness of the shaft and its bearings and restrained torsionally by a twist stiffness of the shaft. This idealization assumes that the gearcase is rigid or can be considered to be part of the support spring and that one end of each shaft is attached to an inertia which is large enough to prevent torsional vibration.

To find the effect of a 1 micron error, i.e., a difference between the displacements x_p and y_p it is simplest to start by assuming that the contact force

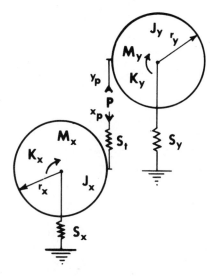

Figure 8.3.   Idealization of a single mesh. The displacement between x_p and y_p is caused by the T.E. and gives rise to a vibration tooth force P.

variation P is known, deducing x_p and y_p and then since $x_p + y_p$ is known, finding P. The total response is obtained by adding the linear and torsional responses so the equations are

$$x_p = \frac{P}{S_t} + \frac{P r^2_x}{K_x - J_x \omega^2} + \frac{P}{S_x - M_x \omega^2}$$

and correspondingly, with no tooth deflection term

$$y_p = \frac{P r^2_y}{K_y - J_y \omega^2} + \frac{P}{S_y - M_y \omega^2}$$

The transmission error $\epsilon = x_p + y_p$, taking note of signs so that from a given ϵ we can deduce P. We can also deduce the forces which act upon the bearing housings; these are given by

$$F_x = P \frac{S_x}{S_x - M_x \omega^2} \quad \text{and} \quad F_y = P \frac{S_y}{S_y - M_y \omega^2}$$

Typical figures in a medium size gearbox might be $r_x = 0.075$, $J_x = .03$ kgm^2, $M_x = 10$ kg, $S_t = 7 \times 10^8$ N/m, $S_x = 10^8$ N/m and $K_x = 5 \times 10^5$ Nm/radian; $r_y = .050$, $J_y = .01$ kgm^2, $M_y = 4$ kg, $S_y = 4 \times 10^8$ N/m and $K_y = 8 \times 10^5$ N/m radian.

Unit force at the mesh at 800 Hz gives deflections of 1.4×10^{-9} due to tooth bend, -21.8×10^{-9} due to rotation effects and -6.5×10^{-9} due to gear center movement giving a total of -26.9×10^{-9} for x_p; the negative sign shows that this part of the system is above a resonance and is moving $180°$ out of phase with the force. The response for the other part of the system is 4.6×10^{-9} for rotation and 3.3×10^{-9} for lateral giving 7.9×10^{-9} in total. The combined response is -19×10^{-9} m/N so that 1 micron T.E. will give -52.6 N at 800 Hz. Bearing forces are $F_x = -52.6 \times 10^8/(10^8 - 10 \times 2^2 \times \pi^2 \times 800^2) = 34.5$ N per micron and $F_y = -70.4$ N per micron.

These estimates can be carried out over a range of frequencies very easily on a programmable calculator and give an indication of trouble areas. The details of each part of the response are interesting since in the calculations the sign of each of the individual terms shows whether the response is above or below resonance and, hence, whether it will be advisable to increase or decrease mass or stiffness to improve response. In the example above the figures obtained should be compared with the tooth force expected at very high speeds well above resonances, i.e., 700 N/micron and the forces which would occur at very low speeds when only stiffness matters and the tooth loads and bearing forces would be 35 N/micron error.

The results obtained from such estimates give the surprising conclusion that torsional resonances are desirable while lateral resonances are not. A torsional resonance, when $K_x = J_x w^2$, gives a very low tooth force for a given error and providing there is no lateral resonance, the bearing forces will be low. There will, however, be high torques transmitted along the shafts and so it is essential to have rotary inertias which will prevent transmission of these torques through the system.

Damping has not been considered and is not known for gearing systems so that it is not worthwhile attempting to predict damping. Estimates of response of systems will give useful results between resonances and will indicate where the resonances are, but only experimental tests will give reliable response information at resonances. Estimates of system masses are usually accurate and the stiffnesses of slender shafts can be assessed accurately but bearings and short shafts give uncertainties and misalignment in a gear mesh shortens the effective face width so that mesh stiffness may vary in less accurate gearboxes.

III. GEARCASE RESPONSE

The estimates in Section II gave the level of forces which would transmit through the bearings to the gearcase. Prediction of the response of a gearcase to excitation can be carried out by finite element techniques but is rarely economically justified because of uncertainties about joint stiffnesses and damping in the system. In designs like car gearboxes, a relatively small gearbox is bolted to an engine block which is supported on antivibration mounts in a chassis and there may well be ten paths in parallel by which gear vibrations can generate noise to reach the passengers. Some noise will transmit directly from gearcase to body panels and some will travel to the engine block and transmit from the block to body panels but most will travel as vibration along the mechanical links between engine and body to excite panels. The complexity of the situation means that only experimental techniques are of use.

In some situations the gearbox is separate and vibration isolated from the power source, e.g., a turbine, and from the load so that it is possible to estimate gearcase vibration levels provided the mounting arrangements are known. Estimates for medium size vibration-isolated gearboxes can be reasonably accurate but should be treated with suspicion until verified experimentally. In the case of both experimental results and theoretical predictions of gearcase response, it is worthwhile expressing the results in terms of the effective mass which the vibrating force sees. In particular, if an oscillating force of 1 N produces a response of 1 m/s^2 the effective mass is 1 kg (or if 10 N produces 1 g acceleration). In a large gearbox forces of 100 N might produce 0.1 g acceleration giving an effective mass of 1 ton (metric). The advantage of specifying system response in this way is that the effective mass of the gearbox can be compared with its actual mass to give a rough measure of the efficiency of the gearcase in preventing vibration reaching its support members.

The gearcase in a machine tool is often the main structure and the dominating source of noise. Modern design styles tend to use large flat panels which are extremely efficient resonators and noise emitters. Conventional case development involves locating troublesome panels with a noisemeter followed by a low mass vibration pickup to identify the mode of vibration then stiffening the panels to break up the patterns and reduce amplitudes. In critical cases damped sandwich panels or extra viscoelastic damping may be required but is expensive. Vibration absorbers may be used for extra damping and are themselves cheap but they are costly to fit as skilled labor is required and they should be set up using a vibrator and pickup to get optimum performance. Complete enclosure of a machine to reduce noise is usually not practicable and can cause heat build-up problems.

IV. TYPES OF NOISE

We have so far considered noise as a T.E. generated vibration which is transmitted through the system until radiated from external panels as noise. It is convenient to Fourier-analyze noise as a collection of steady frequencies and some gearboxes produce a single steady note but, in general, the noise is complex. Frequency analysis of the T.E. or corresponding analysis of the output noise will give information about the type of excitation and, hence, the likely cause or causes.

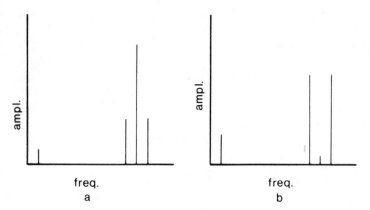

Figure 8.4. Frequency analysis of modulated wave: (a) is for moderate modulation and (b) is for high modulation.

The simplest type of noise is a steady note which may have a harmonic content; frequency analysis will check whether the noise is meshing frequency or an harmonic or in some rare cases a phantom note due to a regular pitch error in machining the gear; deductions are usually then straightforward.

Modulation of a note is very common; instead of a steady note the amplitude or frequency are modulated slightly at a low frequency which is usually, but not necessarily, once per revolution of a shaft. Amplitude modulation is sketched in Chapter 6, Figure 6.6, and we can represent it simply by an expression of the form $\cos\omega t(A + B \cos\Omega t)$ where Ω is once per revolution frequency. Wave analyzers do not recognize unsteady signals and so measure the signal as the sum of $A \cos\omega t + B \cos\omega t \cos\Omega t$ which is $A \cos\omega t + 0.5 B \cos(\omega + \Omega) + 0.5 B \cos(\omega - \Omega)$. This gives the characteristic wave analysis, provided the analyzer has fine resolution, shown in Figure 8.4(a). But, if the modulation is severe, the tooth frequency term disappears almost completely as shown in Figure 8.4(b). Modulation explains why humans "hear" frequencies which it is not theoretically possible for them to detect. A frequency of 10 Hz is inaudible but 10 Hz modulation of tooth frequency of 500 Hz is clearly audible. Modulated noise is often described as a throbbing sound.

Frequency modulation is somewhat similar to amplitude modulation and can be thought of as a regular speeding up and slowing down, usually once per revolution of a gear. The trace looks as shown exaggerated in Figure 8.5(a) and is described mathematically as $A \cos\omega(1 + \epsilon \cos\Omega t)t$; Fourier analysis of this is somewhat complex if the change in frequency is large (3) and gives many sidebands, spaced at $\pm\Omega$, 2Ω, 3Ω, etc., from the carrier (tooth) frequency as in Figure 8.5(b). Gear frequency modulation is usually restricted to small changes in frequency and the effect is then to produce two sidebands at $\pm\Omega$ as with

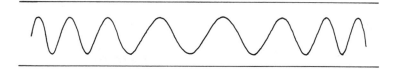

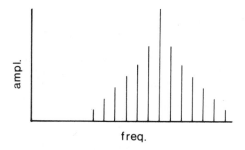

Figure 8.5. Frequency modulation. Amplitude is constant but frequency varies. The corresponding power analysis has a range of sidebands.

amplitude modulation in Figure 8.4(a). To the listener the effect is the same, that of a clear note with a "throb" in it.

There are numerous possible causes of signal modulation and it is not generally possible to say which particular cause has produced the effect. Cyclic variation of torque in the system, for whatever reason, will tend to give an amplitude modulation of the signal, detectable by inspection of the vibration trace. Cyclic pitch errors from an eccentrically mounted or machined gear will give frequency modulation as will also speed variation due to reciprocation or to a cyclically varying load such as a propellor which gives a 4 times per revolution torque variation. Mounting a helical gear with wobble (swash) gives a cyclic variation of noise which is very powerful. In most engineering systems, however, variation of torque and speed go together so that variation of frequency and amplitude occur in the signal, and frequency analysis gives little direct clue to the cause (or causes).

Other types of noise from gearboxes are repetitive but not single or modulated frequencies and harmonics; they are associated with pitch errors on the gears. These noises may be described as scrunching, grating, graunching, grumbling, etc. and contain a very wide range of frequencies when analyzed. The basic once-per-revolution frequency at which the noise repeats is too low to be audible, but the pitch errors generate an irregular series of impulses. A waveform such as that shown in Figure 8.6 is typical of that associated with pitch errors and if played into a loudspeaker will generate a typical gearbox noise. There is,

Figure 8.6. Typical pitch errors. Random pitch errors round a gear will give excitation at all multiples of once per revolution except tooth frequency and its harmonics.

in practice, a waveform repeating from each gear at different speed and the noise heard has no discernable pattern to it but the noise is not white noise and is different in character from white noise despite containing very many frequencies. Although frequency analysis is of little help for this noise, repetitive averaging (see later) will show a clear excitation pattern and indicate the cause of the trouble.

Occasionally white noise will come from a gearbox, with all frequencies present, and will show frequencies that cnnot be generated by the gear errors. This type of noise generation is usually associated with loss of contact between the teeth; when the teeth meet again the impulses which occur generate all frequencies and the output may be very irregular when the loss of contact is irregular.

Pitch errors on a large gear may be randomly distributed and give rise to excitation at unexpected frequencies (4); a gear can give excitation at all multiples of once per revolution and any of these frequencies can pick up a gearcase resonance to give irritating noise. This effect is most likely when tooth frequencies are too high to give resonance troubles but the gearcase has low-damped resonances. Under these circumstances it is necessary to carry out a frequency analysis of the pitch errors on the gear to see if there is strong excitation at a particular harmonic. Regrinding one or two teeth may then be sufficient to eliminate trouble.

Pitch errors can be of use and are sometimes deliberately introduced into gears or toothed belt or chain drives to suppress regular once-per-tooth errors. The pitch errors added produce a background "grumble" but break up the regular whine of tooth frequency harmonics. We do not know whether this is solely due to a reduction in the sound power level at whine frequency or whether there is in addition a masking effect in human hearing which makes the sound less irritating.

Gear location or drive may not be satisfactory in drives where helical gears are mounted on splines or are driven via synchromesh sleeves with spline dog teeth. Any play between gear and support spline will allow irregular movement of the gear on the spline as the tooth contact forces continually reposition the gear on the spline. This will give a noise similar to pitch errors on the gear; misalignment of synchromesh sleeves can give a similar effect.

V. PLASTIC GEARS

At first sight the use of plastics for gears would seem unlikely since steel gears require very high surface hardnesses to achieve a resistance to contact stress pitting. Materials such as nylon or acetal plastics have yield strength of only about 7×10^7 N/m^2 (10,000 psi) and have fatigue strengths which decrease continuously with number of cycles to less than a tenth of their yield strength.

The advantage of plastic lies in the low value of modulus with values of about 3×10^9 N/m^2, about 1.5% that of steel. This allows the contact area to rise greatly, reducing contact stresses, although it does not help root stresses. The result is that contact stresses do not limit gear life and only root stresses need be checked; plastic gears are typically about one tenth the strength of steel gears of the same size for 10 million cycles. Heat input to the plastic is critical and so corrections must be applied for ambient temperature and running speed since these both affect bulk temperature and significant weakening occurs at temperatures well below 100°C.

Since costs of plastic gears are much lower than those of steel gears they are economic for lightly loaded drives; it is quite common to have domestic equipment with plastic gears at the high speed motor (20,000 rpm) followed by steel gears at the low speed output. Expansion is ten times that of steel and moisture can give 1% size increase so backlash must be high. Lubrication is not essential but improves load carrying capacity as it helps cooling and reduces heat input to the plastic. It is advisable to approach the plastic manufacturers when design of gears is required.

From the aspect of noise the low modulus is a big advantage since the mesh stiffness is reduced by a factor of 70 for the same facewidth. Moulded or cut plastic gears are surprisingly accurate and if we assume errors about 2.5 times as large as those of steel gears, the reduction in forces above system resonances is of the order of 30, i.e., 30 dB down on noise. Below resonances the vibration levels will be higher with plastic than with steel; as the mesh stiffnesses are lowered the resonance frequencies of the system will be reduced but this is not a major effect.

REFERENCES

1. Ishida K., On the friction noise and pitch circle impulse noise of gears. J.S.M.E., 1967, Semi-International Symposium, 4-8, Sept. 1967, Paper 318, pp. 161-170.
2. Aitchisin T. W., Correlation of gear noise with manufacturing process. N.E.L. Report No. 507, April 1972, N.E.L., East Kilbride, Glasgow.
3. Thompson A. M., Fourier analysis of gear errors, NELEX 80, October 1980, N.E.L. Glasgow, Paper 3.5 (available from Gleason-Goulder).
4. Welbourn D. B., Gear errors and their resultant noise spectra. I. Mech. E. Gearing in 1970 Conference, pp. 131-139.

Chapter 9

EXPERIMENTAL INVESTIGATIONS

I. FREQUENCY ANALYSIS OF NOISE

The most widely used technique for analysis of noise or vibration from a gearbox is frequency analysis. In some cases it is unnecessary because viewing the waveform on an oscilloscope is sufficient for pure regular notes. In general the signal is confused or irregular and so some form of filtering or analysis is required.

Analysis may be termed frequency, wave, Fourier, Fast Fourier, power, spectral, power density, spectral density, harmonic, real time, third octave, etc. These terms are used indiscriminately and often interchangeably; most of them mean the same thing and all are concerned with expressing the amount of noise at a particular frequency or in a particular band of frequencies. The mechanism by which the analysis is carried out is not important; the older approach usually termed wave analysis uses a filter (analog) which is tuned to a particular frequency and is slow in operation. Modern methods tend to be digital and, hence, much faster in operation but work with a limited number of samples; digital methods can give much narrower frequency bands than analog methods and so are useful to discriminate sidebands.

Typical analysis filter bands are shown in Figure 9.1 which gives an idea of the bandwidths that would be obtained at 250 Hz from a typical 2% wave analysis and a 400 line digital analyzer set to 500 Hz compared with an ideal filter. Digital methods have virtually no limit provided enough samples of the vibration are taken; in practice the resolution is limited by the finite number of samples (typically about 2000) which are used for the computation.

The important difference between the two "families" of instruments lies in whether they measure wave amplitude at a particular frequency or the amount of power in a given frequency band; the units in one case are volts, representing sound pressure level or acceleration, whereas in the other case the units are volts squared per Hertz. The power spectral density is what is usually measured (in g^2 per Hz) for random processes and is, unfortunately, not what is required for gearbox work. A truly random process occurs over a wide range of

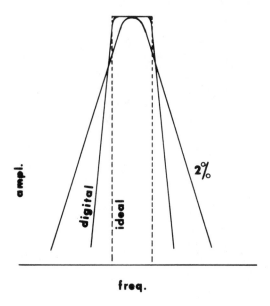

Figure 9.1.   Sketch of filter characteristics.

frequencies and what is required is the power in a narrow frequency band because this remains reasonably constant whereas the power at a single frequency
is theoretically zero. The plot of power density against frequency will be as in
Figure 9.2 and will be unaffected by the width of analysis band chosen since
halving the width of the band will halve the power measured. In contrast the
noise generation from a gearbox can theoretically only occur at frequencies
which are multiples of once per revolution and the power occurs at exact frequencies; this type of noise or vibration output is called a line spectrum and a
power density analysis would appear as in Figure 9.3. In this case there is a
certain amount of power in the signal at a given exact frequency, e.g., there may
be 3.5 microns T.E., 0.75 g acceleration or 84 dB (A) sound at a frequency of
501.5 Hz. A wave analyzer will read this amplitude (assuming no neighboring
sidebands) regardless of whether it is set on a narrow bandwidth or a wide one,
whereas a power spectral density analysis will give a larger answer the narrower
the bandwidth. For the figure given the spectral density obtained would be
3.5^2 microns^2/Hz if the bandwidth were 1 Hz but proportionately higher or
lower if the bandwidth were less or greater.

   Conversion from power density to amplitude of vibration when a line
spectrum is involved requires that the V^2/Hz figure from the analyzer is multiplied by its bandwidth to give volts squared, i.e., power and then the square
root taken to give the actual voltage. As an example, if the output is 0.6 V^2/Hz,

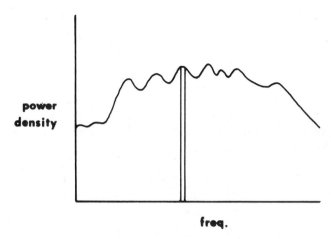

Figure 9.2. Random process power density. Power exists at all frequencies.

the bandwidth is 2.5 Hz and the transducer sensitivity is 500 mV/g what is the amplitude of vibration at 320 Hz? The power involved in this band, all due to one line frequency, is 0.6×2.5 which is 1.5 volts^2 and so acceleration is $1.225/0.5 = 2.45$ g $= 24.03$ m/s^2 (r.m.s.). This acceleration at 320 Hz corresponds to an amplitude of vibration of $24.03/(2\pi \times 320)^2 = 5.9$ microns r.m.s. which is 16.7 microns peak-to-peak.

Frequency analysis of regular repetitive waveforms such as those from a particular gear gives an answer which sometimes disagrees with common sense.

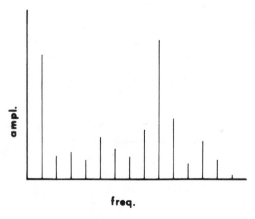

Figure 9.3. Gearbox power density spectrum. Only multiples of once per revolution (or once per mesh cycle) are possible.

Figure 9.4. Repetitive waveform. If this waveform repeats at once per revolution of a shaft the frequency ω will not appear in a power analysis unless it is an
exact multiple of once per revolution.

If we take the waveform shown in Figure 9.4 and play it repeatedly into a waveform analyzer with a narrow bandwidth the result will not necessarily show a
high value at the observed frequency ω despite the obvious presence of a vibration at this frequency. This anomaly occurs because, as far as the mathematics
are concerned, a waveform that repeats at once per revolution can only contain
frequencies that are multiples of once per revolution and so frequencies such as
ω cannot exist. This contradiction between the mathematics and common sense
is not usually a problem but it should be recognized that frequencies obtained
by Fourier Analysis are not necessarily "real" frequencies.

II. CORRELATION AND WHITEWASHING

The signal from a gearbox, whether vibration or sound pressure level, is often
confused and "noisy." It is rather muddling that "noise" is used in two completely different senses; in one case noise refers to sound transmitting through
the air whereas in the other case it means any signal (usually electrical) which
comes from any source other than the gear mesh of interest.
     Correlation is a mathematical technique which has the effect of extracting
any regular signals from an apparently random vibration. It is defined (1) as the
sum of the products of the level of a signal at one time multiplied by the level of
a signal time τ later. In the particular case of autocorrelation a signal level is multiplied by the level of the same signal time τ later and all such products are
summed. This operation has similarities to Fourier analysis in that if there is any
regular waveform repeating with interval τ the products consistently add whereas
all other waveforms, whether random or repeating at different frequencies, give
both positive and negative products which cancel when summed. Figure 9.5
shows a regular signal and shows the corresponding autocorrelation function.
     In the particular case of a basic sine wave of period τ submerged in a
"noisy" random vibration, autocorrelation of the signal eliminates the random
or uncoordinated part of the vibration and leaves the sine wave as a clean wave
without random noise; several sine waves of different frequencies will still appear
in the final signal provided they occur regularly so the effect of autocorrelation

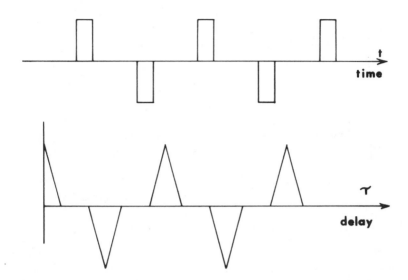

Figure 9.5. Correlation. Repetitive pulse signal and corresponding autocorrelation function obtained by taking each signal value and multiplying by the signal value time τ later.

is a "cleaning" of the signal. Because of this it is sometimes called "whitewashing."

The usefulness of correlation arises from the fact that phase information still exists after correlation whereas frequency analysis does not display phase information. A single once-per-revolution short pulse contains all harmonics of once per revolution; frequency analysis does not tell us whether the harmonics are phased to give an impulse or whether they are phased so that the vibration is spread throughout the revolution and looks like white noise. Figures 9.6(a) and 9.6(b) are two repeating signals whose frequency analysis may look the same whereas autocorrelation will give very different results.

Cross-correlation between the output and input of a vibration transmission system can sometimes be of use in identifying pulse transmission times and, hence, paths in large structures but is rarely of use in gearboxes because we do not usually have the input, i.e., the T.E. available. Autocorrelation of the output signal, i.e., the noise can be useful for identification of which shafts contribute the most noise but is little used since averaging techniques are better.

In the particular case where a once-per-revolution pulse is available, cross-correlation of the output signal with the pulse gives an output which is solely the part of the noise which repeats at shaft frequency. This technique gives the same answer as averaging if the pulse is short but is slow and wasteful of computational effort since a large number of irrelevant multiplications are needed so that it is not used.

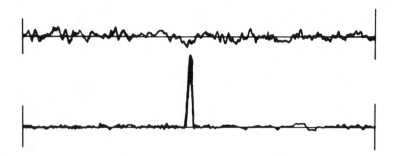

Figure 9.6.   Repeating signals. Both signals contain power at a range of frequencies and give similar power density spectra but will differ with autocorrelation.

III. AVERAGING TECHNIQUES

In a complex gearbox, frequency analysis is useful to identify which tooth mesh frequencies are giving trouble and it is possible, by studying the harmonics of once per revolution and the sidebands either side of tooth frequency, to get an idea of possible causes of trouble.

The problem with this approach are that a given set of sidebands or harmonics can have more than one interpretation and often the only useful information from a frequency analysis is identification of tooth frequency which is already known. When a gearbox has pitching errors which give a "grumbling" sound and power at all possible frequencies the power distribution is of little help.

Signal averaging is a technique which is coming into more general use, particularly for gearbox monitoring (2), (3), as it is the best method for detecting small variations in the vibration pattern long before power frequency analysis techniques show up trouble. It is necessary to have an accurate once-per-revolution marker to trigger successive sweeps of an averager which at each of, say, 1024 positions round the revolution adds the level of vibration measured for perhaps 128 revolutions and displays the summed result. The resulting average contains negligible contributions from any random vibrations and from all other shafts which are not running at the same speed and so represents the average contribution to the vibration (or sound) from that shaft. This averaging process is extremely fast and requires little computing capacity and so it can be carried out quickly and cheaply on line even when speeds and frequencies are high; 100 kHz sampling rates are easily achieved.

The sensitivity of the technique can easiest be appreciated by considering the waveform shown in Figure 9.7 where there is slightly irregular pitching, a regular once-per-tooth disturbance with harmonics and a single slightly damaged tooth with damage of about the same amplitude as the regular tooth disturbance.

Figure 9.7.   Waveform with tooth damage on one tooth.

This damage stands out visually and would be detected as a peak to r.m.s. value
of about 3 using averaging techniques. The vibration power involved however is
small because, although the signal is double amplitude at the trouble spot, it is
only occurring for 1% of the total revolution (if there are 50 teeth) so that the
increase of power level at once-per-tooth frequency is about 30 dB down on the
steady power level from the other teeth and so is undetectable. The effects on
the other harmonics are also small and the presence of the damage may reduce
rather than increase the power observed at harmonics of once per revolution.

The general use of averaging techniques in gearbox work is likely to increase because the form of the output gives a clear understandable picture of
what is occurring. Figure 9.8 is the T.E. of a two mesh standard in line car gearbox and gives an idea of the irregular beating effects that are observed whereas
Figure 9.9 is averaged traces taken under the same conditions and shows clearly
the regularity of the manufacturing eccentricity and pitch errors as well as the
effects of load on the mesh alignment giving large variations in the once-per-
tooth component.

Once-per-tooth errors in a particular mesh are detected by averaging but
care is needed in deductions because the 1/tooth error in a mesh usually consists
of a "steady" component upon which is superimposed the variations from the
average of both gears. Averaging at either shaft frequency shows the steady component in both cases with the relevant unsteady component superimposed. It is
not possible to allocate blame for the "steady" component to either gear because lack of conformity is a joint problem; it is a purely arbitrary choice
whether one gear is considered "right" and the other "wrong" since whether a
designer chose, say, a helix angle of $28°50'$ or $28°40'$ was arbitrary.

A particular case in which frequency analysis and averaging work well together occurs when modulation of tooth frequency (or harmonics) is occurring.
Amplitude and frequency modulation both give sidebands in the frequency
analysis (see Chapter 8, Section IV) but, whereas frequency modulation does not
show clearly on an averaged trace, amplitude modulation does. It is thus possible
to measure the total power in the sidebands, subtract the amplitude modulated
part deduced from the averager trace and deduce the amount of frequency
modulation. This helps to allocate blame to alignment errors, to system
dynamics or pitching errors.

The disadvantage of averaging is that an extra signal is required from an
accurate once per revolution marker. It is not accurate enough to get the sweep

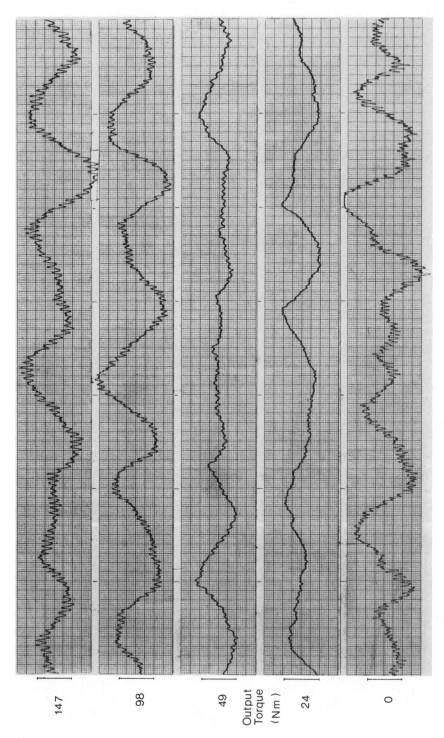

147

98

49

Output
Torque
(Nm)

24

0

Figure 9.8. T.E. measurements from a twin mesh gearbox. 48 rpm. Scale marker 10 arc minutes of input.

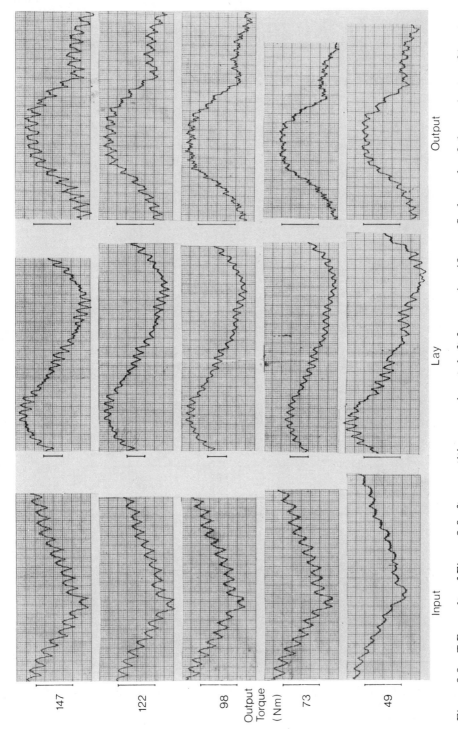

Input   Lay   Output

147   122   98   Output Torque (Nm)   73   49

Figure 9.9.   T.E. results of Figure 9.8 after repetitive averaging at shaft frequencies. 48 rpm. Scale marker 3.6 arc minutes of input.

speed roughly right since even 0.1% speed inaccuracy on the sweep would give garbage results if more than about 8 revolutions are averaged. "Jitter" is the other hazard which can affect results when speed varies from revolution to revolution. 1% jitter on a 50 tooth gear will give 180° phase change by the end of the revolution so the results will be useless towards the end of the sweep. This problem can be overcome by fitting extra sampling pulse generators giving say 60 times per revolution pulses but this represents an experimental complication.

A slight advantage of averaging arises from its reduced digital memory requirements since only 1024 bits of information need be stored at any one time whereas a complete mesh cycle on a standard car gearbox takes 600 revolutions, i.e., some 10 secs and involves sampling at least half a million data points per channel.

IV. SPEED VARIATION

Frequency analysis of vibration or noise from a gearbox gives a plot which may be typically as shown in Figure 9.10(a). There will be several resonance peaks, some of which are easily identified as expected forcing frequencies such as tooth frequency and some of which have no apparent cause but might be due to

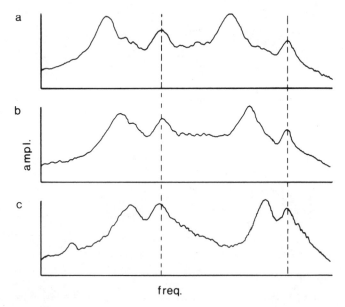

Figure 9.10. Variation of power spectrum with speed. Increase of speed from (a) to (b) to (c) moves two peaks up in frequency but the other two peaks remain at fixed frequency.

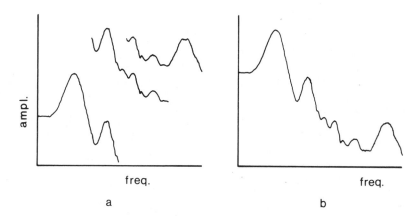

freq.   freq.

        a   b

Figure 9.11.   System response. Variation of the speed with filters set at, say, tooth frequency, 2nd harmonic and 3rd harmonic gives the sections of the response curve in (a) and these can be aligned to give the single curve in (b).

harmonics of once per revolution due to pitch errors. A plot such as this at a single speed does not tell us whether a particular peak observed is due to high excitation with low gearbox response or normal excitation coinciding with a resonance of gearbox internals or casing. Increasing speed as in Figures 9.10(b) and 9.10(c) will increase excitation frequencies but not affect resonance frequencies and we can deduce that peaks which move up the frequency scale are associated with forcing whereas fixed frequency peaks are associated with system resonances. When the two coincide extra high peaks are obtained. This gives a qualitative measure of which resonances matter and where they are.

   This approach can be taken a stage further by varying speed, preferably at constant torque, over as wide a range as possible and using filters to extract parts of the response. If we set a filter to tooth frequency and run through the speed range we know that the excitation remains constant since it is always the tooth frequency component of the T.E.; the output of the filter will then give the response of the system as input frequency varies but amplitude does not, as in a standard vibration test. The filter output gives the shape of the gear system response regardless of the particular set of gears fitted. Repeating this technique at other forcing frequencies gives a series of overlapping sections of the response as in Figure 9.11(a) and these can be lined up to give a composite response as in Figure 9.11(b).

   We can also set a filter at a fixed frequency, preferably on a resonance, and vary speed; in this case the system response is fixed and the excitation at this frequency varies with speed. The corresponding plots at different fixed frequencies are as shown in Figure 9.12(a) and are superposed in Figure 9.12(b). Response is

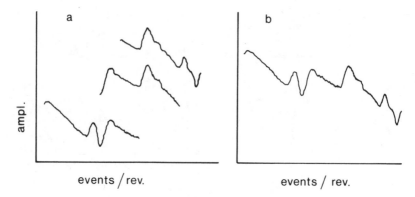

Figure 9.12.   System excitation. Variation of speed with filters at fixed (resonance) frequencies gives the relative amounts of excitation as number of cycles per revolution varies. The sections of curve in (a), each obtained at a different frequency, may be overlapped as in (b) to give the relative sizes of the components of the excitation, i.e., the T.E.

plotted against number of events occuring per revolution so that if a filter (resonance) is at 1200 Hz and the speed is 1800 rpm the output corresponds to 40 excitations per revolution; doubling speed gives 60 revolutions per second and so gives the output corresponding to 20 excitations per revolution.

These two sets of plots in Figures 9.11(b) and 9.12(b) then give a clear picture of the relative response of the resonances and the relative amplitudes of excitation in the system; they do not however give absolute values of meshing error unless T.E. is being measured dynamically. Occasionally in lightly loaded or inaccurate gearboxes the individual sections of curves do not agree in shape where they overlap; this indicates that non-linearities are occurring and the basic assumed linear theory can no longer be applied.

The same information can be displayed as a "waterfall" plot (4) which shows the way the fixed and moving peaks vary throughout the range; this gives a very quick overall picture of the system behavior.

When speed cannot be varied continuously it is more difficult to analyze the results. Research by Daly (5) on the problem of deducing inputs and system response from tests at a limited number of speeds suggested that 3 different speeds would give reasonable results provided that they were not too close together.

Test results from a gear drive with helical gears are given by Kohler, Pratt, and Thompson (6) who compare outputs with T.E. and show reasonably consistent system responses over a range of running speeds; T.E. was measured directly at speed using torsional vibration transducers. A digital approach was used to give the line spectra for input and output at a range of speeds and corresponding lines were compared directly to give system response.

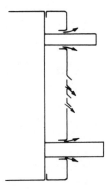

Figure 9.13. Sound leakage. Noise from a cover may either come from transmission through gaps as indicated by the arrows or may be due to the cover itself vibrating.

V. CASING AND STRUCTURE MEASUREMENTS

There are no general rules for testing and developing gearbox casings since they vary so much that what is relevant in one investigation is misleading in another.

The simplest cases occur when a single frequency (usually tooth) dominates the sound and so only this frequency need be considered. Changing internals is usually more costly than altering a casing so the casing is attacked first.

A noisemeter scan round the outside of the gearbox while it is running is usually worthwhile but is absolutely essential if there are vented guards as shown diagrammatically in Figure 9.13 since only local sound measurement will give an

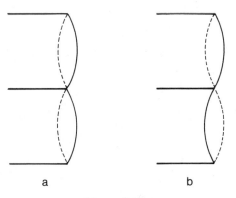

a   b

Figure 9.14.

idea of whether the sound is mainly coming out of the vents or is being transmitted by the guard. After a preliminary check with a noisemeter it is best to turn to a velocity pickup since this is not influenced by standing wave effects or external noise.

Velocity is most convenient in practice because acceleration overemphasizes the high frequency content and displacement, the low frequency, once per revolution content of the vibration. It is usually obtained by using a lightweight accelerometer and integrating the output to give velocity. The measurements of velocity give the mode shape of the system and indicate the trouble areas; it is worthwhile having a reference pickup to check phase of the vibration since a pair of panels vibrating as in Figure 9.14(a) is a very much more powerful sound source than a pair vibrating in antiphase as in Figure 9.14(b).

Some gearboxes are as sketched in Figure 9.15(a) with fairly flat sides which are rigid in the direction of the side but flexible in bending. A mode shape as measured in 9.15(b) with center amplitudes much larger than those at the sides suggests a panel resonance that should be eliminated, whereas a mode shape as in Figure 9.15(c) suggests that the panel is already above its resonant frequency and stiffening would increase sound levels. Panels may sometimes be isolated from the vibration of the main structure by rubber mounting but only non-structural covers, etc., can be tackled this way.

In some designs it is found that as many as 20 different panels, virtually the whole of the casing, are vibrating with so many modes that the overall effect is that of white noise emission from the whole of the surface. Detailed tinkering with design by adding a few stiffening ribs is unlikely to have much effect in such a case so either large overall increases in case thickness are required or change to a highly damped material is necessary. If neither of these is possible an isolation of the bearing housings from the outer case must be attempted.

Structure borne vibrations are rather more difficult to assess and require more careful measurements. In designs where there are vibration isolators it is possible to make measurements either side of the isolator, taking careful note of

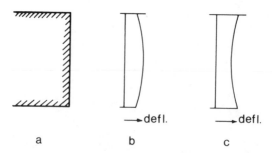

Figure 9.15. Panel mode shapes. (a) represents the side of a gearcase and (b) is a mode shape with amplification of edge support amplitudes wheareas in (c) the center amplitudes are less than the edge amplitudes.

Figure 9.16.

phase, and then calibrate the isolator in a dynamic rig to allow deduction of the force passing through the isolator. Even better, when it is possible, is insertion of a force transducer under the feet of the gearbox; piezo-electric washers may be used when the direction of the force is known. In general when the vibration pattern is not known it is better to use 6-component "top-hat" (7) dynamometers which are extremely simple and rigid. Figure 9.16 shows a dynamometer in position supporting a rear suspension assembly. The active section is a thin cylindrical shell which is strain-gauged to give local vertical and shear stresses; these stresses, suitably added or subtracted, give the 3 components of force and 3 moments. Modern electronics allow measurements of forces of the order of 1 N with a dynamometer which can withstand 10 kN (1 ton). Larger and smaller dynamometers have been designed to take 100 Tons and to measure fingertip forces down to 0.1 N with torques of 10^{-3} Nm.

REFERENCES

1. Newland D. E., Random vibrations and spectral analysis. Longmans, London and New York, 1975.
2. Stewart R. M., Vibration analysis as an aid to the detection and diagnosis of faults in rotating machinery. Inst. Mech. Eng., Vibrations in Rotating Machinery Conference, Sept. 76, Cambridge, pp. 223-229.
3. Daly K. J. and Smith J. D., Using gratings in driveline noise problems. Inst. Mech. Eng. Noise and vibrations of engines and transmission Conference, Cranfield, July 1979, pp. 15-20.
4. Keller A. C., Advantages of high speed and real time F.F T. analysers, Sound and Vibration, March 1978 (available from Spectral Dynamics).
5. Daly K. J. and Smith J. D., Estimation of excitation and transmissibility from output measurements with application to gear drives. Journal of Sound and Vibration, 1981, 75 (1), pp. 37-50.
6. Kohler, H. K., Pratt A., Thompson A. M., Dynamics and Noise of parallel axis gearing. Proc. Inst. Mech. Eng., Vol. 184 (69-70), Pt 3, O, pp. 111-121 (Conference Gearing 1970).
7. Smith J. D. and Welbourn D. B., A six component dynamometer. Journal of Mechanical Engineering Science, April 1970, Vol. 12, No. 2, pp. 143-145.

Chapter 10

GEARBOX MODELLING

I. COMPLEX SYSTEMS

In Chapter 8 a basic lumped parameter model of a single mesh had four degrees of freedom and gave contact forces and bearing forces for chosen T.E. frequencies. This simple approach soon breaks down as the model becomes more complex because either the number of degrees of freedom becomes too large, or the system loses contact and becomes non-linear or the system is "large" and no longer behaves as an assembly of lumped masses.

For linear systems in which the main requirement is to know the natural frequencies and mode shapes the most popular approach is to use a matrix method. The amount of work required to describe a system is large and care is needed; an example of a two-dimensional epicyclic gear is typical:

Figure 10.1 shows one planet in an epicyclic gearbox and we wish to write down the equation of motion in the tangential direction. Selecting co-ordinates is an arbitrary choice with little difference between using radial and tangential or cartesian coordinates for the planet center movements; A represents an angle measured counterclockwise in all cases and T is correspondingly measured counterclockwise. Then

$$M_p \ddot{T} = + K_p(-T + V_c \cos Q - H_c \sin Q + cA_c)$$

$$+ K_t \cos \phi(-T \cos \phi + V_s \cos \overline{Q + \phi} - H_s \sin \overline{Q + \phi} + sA_s + pA_p + R \sin \phi)$$

$$+ K_t \cos \phi(-T \cos \phi - pA_p - R \sin\phi + V_a \cos \overline{Q - \phi} - H_a \sin \overline{Q - \phi} + aA_a)$$

Provided the contact stiffnesses are equal as has been assumed in this case, certain of the terms cancel and the equation becomes:

$$+K_t \cos \phi \, aA_a + K_p \, cA_c + K_t \cos \phi \, sA_s - K_t \cos \phi \sin \overline{Q - \phi} \, H_a$$

$$- K_t \cos \phi \sin \times \overline{Q + \phi} \, H_s - K_p \sin Q \, H_c + K_t \cos \phi \cos \overline{Q - \phi} \, V_a$$

$$+ K_p \cos Q \, V_c + K_t \cos \phi \cos \times \overline{Q + \phi} \, V_s - (K_p + 2 K_t \cos^2 \phi) T + M_p w^2 T = 0$$

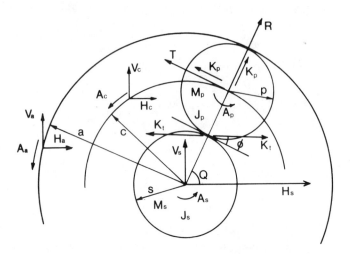

Figure 10.1. Coordinate system for sun, annulus planet carrier, and one planet in a 2-D epicyclic. V, H, T, and R are linear displacements; A represents an angle; K is a support or contact stiffness; M and J are inertia and corresponding moment of inertia.

The corresponding torsional equation is simpler and reduces to:

$$K_t(-2pA_p + V_a \cos \overline{Q - \phi} - H_a \sin \overline{Q - \phi} + aA_a - 2R \sin \phi - V_s \cos \overline{Q + \phi}$$
$$+ H_s \sin \overline{Q + \phi} + sA_s) + J_p \, w^2 \, \frac{pA_p}{p^2} = 0$$

In the original equation pA_p, aA_a, etc. always occur together so it is simplest to regard them as the variables and the effective masses are J_p/p^2, J_s/s^2, etc. The equations can then be written out in the matrix form as in Figure 10.2.

Figure 10.2. Matrix form of equations for 2-D epicyclic. S_{ij} is force at i for unit displacement at j and $s_{ij} = s_{ji}$. Frequencies and mode shapes are given by $SX = \omega^2 MX$ and the values of stiffnesses are:

TT	$= -K_p - 2K_t\cos^2\phi$		RR	$= K_p - 2K_t \sin^2\phi$
HT	$= -K_t \sin(Q+\phi)\cos\phi$		HR	$= K_t \sin(Q+\phi)\sin\phi$
A_sT	$= -K_t \cos\phi$		A_sR	$= K_t \sin\phi$
VT	$= K_t \cos(Q+\phi)\cos\phi$		VR	$= K_t\cos(Q+\phi)\sin\phi$
A_sA_s	$= -\Sigma K_t$		A_pR	$= -2K_t \sin\phi$
A_pA_p	$= -2K_t$		A_pA_s	$= K_t$
A_pH	$= K_t\sin(Q+\phi)$		A_pV	$= -K_t\cos(Q+\phi)$
HH	$= -K_s-\Sigma K_t\sin^2(Q+\phi)$		VV	$= -K_s - \Sigma K_t \cos^2(Q+\phi)$

$$
\begin{bmatrix}
S_{TT} & O & O & S_{TH} & S_{TA_S} & S_{TV} \\
O & S_{A_pA_p} & S_{A_pR} & S_{A_pH} & S_{A_pA_S} & S_{A_pV} \\
O & O & S_{RR} & S_{RH} & S_{RA_S} & S_{RV} \\
S_{HT} & S_{HA_p} & S_{HR} & S_{HH} & O & O \\
S_{A_ST} & S_{A_SA_p} & S_{A_SR} & O & S_{A_SA_S} & O \\
S_{VT} & S_{VA_p} & S_{VR} & O & O & S_{VV}
\end{bmatrix}
\begin{bmatrix}
T_1 \\
pA_p \\
R_1 \\
H_S \\
sA_S \\
V_S
\end{bmatrix}
$$

$$
= \omega^2
\begin{bmatrix}
M_p & & & & & \\
& \dfrac{J_p}{p^2} & & & & \\
& & M_p & & & \\
& & & M_S & & \\
& & & & \dfrac{J_S}{s^2} & \\
& & & & & M_S
\end{bmatrix}
\begin{bmatrix}
T_1 \\
pA_p \\
R_1 \\
H_S \\
sA_S \\
V_S
\end{bmatrix}
$$

There will be 18 coordinates and, hence, 18 equations for a 3 planet design and 24 for a five planet design. There are zeros in the stiffness matrix wherever elements are not connected directly, i.e., all terms linking planets are zero as are carrier to annulus or carrier to sun; there are also some terms which are zero by symmetry. Any term such as the Hs to Vs link has the sum of sin $\overline{Q + \phi}$ terms from each planet and will sum to zero for a symmetrical box as will also the torsional to linear links for sun, carrier, and annulus. The summation terms such as the $- \xi \, kt \sin^2 \overline{Q + \phi}$ for the restoring effects of the planet contact on horizontal displacement of the sun simplify in practice to Kt times half the number of planets. If the design is such that the annulus or planet carrier does not move then the stiffness matrix is correspondingly reduced and simplified.

Once the information is in matrix form standard routines will give the eigenvalues, i.e., the resonant frequencies and the eigenvectors, i.e., the mode shapes (1).

A simpler problem, that corresponding to Figure 8.3 gives a 4 × 4 matrix which gives a stiffness matrix as:

$$
\left|
\begin{array}{cccc}
-(S_x + S_t) & S_t & S_t & S_t \\[2mm]
S_t & -\left(S_t + \dfrac{K_x}{r^2}\right) & -S_t & S_t \\[2mm]
S_t & -S_t & -(S_y + S_t) & S_t \\[2mm]
S_t & S_t & S_t & -\left(S_t + \dfrac{K_y}{r^2}\right)
\end{array}
\right|
\quad
\begin{array}{c}
x \\[2mm]
rA_x \\[2mm]
y \\[2mm]
rA_y
\end{array}
$$

In each case r corresponds to the associated x or y radius and the effective inertias are M_x, J_x/r^2, M_y, and J_y/r^2.

Inspection of the mode shapes to see if there is significant relative deflection at the teeth is sufficient to see whether once-per-tooth errors can feed in energy to excite a resonance strongly or not. Calculation of the response (undamped) can be done by assuming a force at the teeth as in Chapter 8, Section II then summing deflections to equal the T.E.

The matrix approach is very useful for identifying resonances but is of no help in predicting resonance amplitudes, mainly because damping is unknown. It should give reliable information between resonances and is sufficient for preliminary work where there is little existing information about the system. It is dependent on linearity in the system and so cannot be used if there is any danger of loss of contact at the teeth or significant non-linearity in the bearings and the system must then be modelled in detail.

It has been assumed so far that the casing is relatively rigid; this is a reasonable assumption for a machine tool gearbox but not for a gearbox in an airplane because weight reduction gives flexible casings. Analysis of the system dynamics is then very much more complicated. Finite element techniques are used to predict the casing characteristics and can be used to predict resulting vibrations as described by Drago (2) but there is interaction between casing response and internals so that the whole system should be modelled using finite elements. This is expensive since so much effort is involved and so can only usually be done for government financed work; for commercial work it is more economic to carry out simple predictions than develop experimentally.

II. NON-LINEAR SYSTEMS

A non-linear system has "resonances" which are no longer at fixed frequencies but may spread over a two to one frequency range and a regular input does not necessarily give a regular output. There is also the problem that excitation at a particular frequency may give output noise which is not only at harmonics, as expected, but at sub-harmonics; super-position no longer applies so we cannot use a frequency analysis approach with steady sinusoidal inputs.

Digital or analog computing methods may be used to model a gear system and as there is little to choose between them the decision is likely to depend on local facilities. Digital methods are good for generating a range of inputs very easily whereas analog methods deal with loss of contact easily and give the output in a convenient form. The flow chart for the computation is exactly the same in both cases; for each lumped mass the tooth forces and bearing forces are added (with due regard for direction) and divided by effective mass to give acceleration which is then integrated to give velocity and again to give displacement. The tooth forces are obtained by adding the components (up to 6) which contribute to the deflection at a mesh then deducing the force from the amount of interference between the teeth multiplied by the mesh stiffness. When the interference is negative the force is zero. Figure 10.3 shows the idealization of a simple in-line car gearbox and Figure 10.4 is the corresponding flow chart.

In the digital approach the integration is carried out on a step-by-step basis taking a convenient small increment of time to give sufficient frequency response in the output. Assuming that the response is required up to 10 kHz the increment of time should be 25 microseconds or less. Six degrees of freedom re-require that 12 values of velocity and displacement are stored at each instant and the computation should be ideally for a complete mesh cycle, about 10 seconds. This requires nearly half a million stages of computation with some hundred calculations to be carried out at each stage and output storage for half a million points for each variable. Small computers cannot deal with this type of problem economically.

It is a great advantage to be able to listen to the "output" for identification of noise character, as mentioned in Chapter 8. Digital computers do not

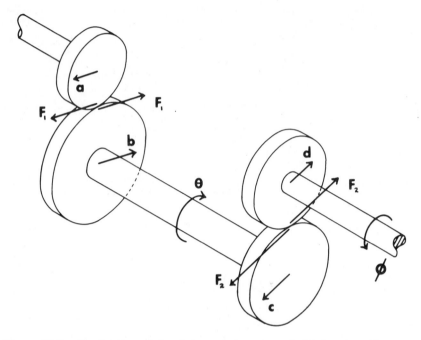

Figure 10.3.   Idealization of simple two mesh system. Axial forces and interactions between (b) and (c) or (a) and (d) are ignored.

normally compute as fast as real time but the output can be speeded up by tape recorder and played into a loudspeaker.

Analog computation has the advantage that it can work in real time so that it is possible to listen to the output directly. Commercial computers are usually designed for work at relatively low frequencies, below 100 Hz, but as none of the normal hold or reset facilities are required it is simple and cheap to mount standard amplifiers costing less than $2 each in a box and connect up using 4 mm plug-in components on the front panel and the system can then work happily above 20 kHz using amplifiers such as the CA 3140, particularly if impedances are kept low. Contact loss is simulated using diodes; an installation to model an epicyclic with 15 degrees of freedom and 6 non-linear tooth contacts requires about 60 amplifiers.

Generating the input T.E. is difficult and digital means must be used to achieve a waveform which is the same shape but whose sweep time can be altered. The simplest method of generating the input is to use a device such as a transient recorder with sampling rate controlled by an external oscillator. The required T.E. waveform can be put in very slowly by hand then replayed repetitively at speed; storage of waveforms in a tape recorder will ensure a consistent

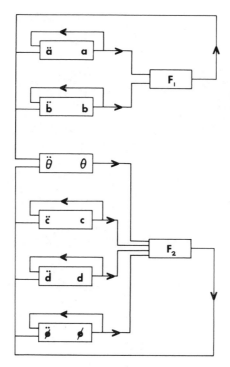

Figure 10.4. Flow chart for system of Figure 10.3.

input. In the case of an epicyclic gear, depending on the numbers of teeth, the once-per-tooth excitations can be in-phase or staggered. Both systems should be checked since in-phase meshing gives strong torsional excitations to sun and annulus whereas phased meshing gives annulus distortion in the form of a travelling wave. In the case of epicyclic planets the choice is between strong tangential forcing of the planet and strong torsional forcing according to the number of teeth.

Damping presents a problem, as usual, for either digital or analog approach and it is usual to provide a small amount of damping of the order of 2% at bearings and at teeth contacts. In the analog case this is done at the teeth by using diodes and slight phase advance as shown in Figure 10.5(a) to give a typical characteristic as in Figure 10.5(b).

Lightly loaded gears (see Chapter 14) spend much of their time out of contact and may bounce right across the backlash to impact on the other side. This introduces extra complexity but is dealt with easily using either approach; a separate T.E. input is however required for the trailing flank mesh since it can either assist or reduce vibration according to the phasing.

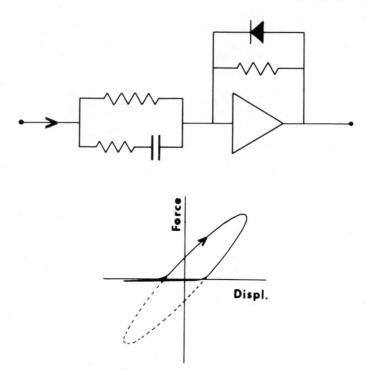

Figure 10.5. (a) Circuit used to simulate tooth contact forces with damping and (b) Corresponding force-displacement relation at high frequency. There is some force developed before contact occurs.

A further possible complication is whether or not to include variable tooth stiffness effects in the model. A high-precision well-designed helical gear will have a contact stiffness which rises approximately 100% from light load to full load but should not vary with time whereas a spur gear mesh stiffness varies principally with time. Time varying stiffnesses give powerful half frequency forcing and can give rise to damage on alternate teeth but are not common. Unless there are special reasons to suspect sub-harmonic resonance in a spur box with light damping it is sufficient to use an average value of stiffness.

III. SYSTEM T.E. INPUT

Prediction or investigation of gearbox response requires an input T.E. regardless of the method of modelling. The ideal is to have an existing gearbox available and to be able to measure T.E. at full torque but at low speeds so that dynamic effects are unimportant. This would give the quasi-static T.E. which is the excitation of the system but it is extremely rare that this is possible; even when a gearbox exists and is available it is not possible to run it slowly at full torque because teeth will scuff or plain bearings will be damaged.

T.E. under load can be measured statically by applying full torque to the system with levers or a slave gearbox, measuring angular position of input and output shafts with a precision optical head or theodolite then moving the input through a small angle corresponding to about a tenth of a tooth and repeating the measurement. This gives exactly the information required but is extremely laborious and it is not easy to apply sufficient torque to very large gearboxes where 100 tons force is required at 1.5 m radius. Accuracy of better than 1 second of arc is difficult to achieve consistently and this corresponds to 2.5 microns at 0.5 m radius so that only small or medium size gearboxes may be tackled this way. If the system uses plain bearings care is needed because there will be extra static deflections at the bearings of the order of 25 microns leading to extra tilt for overhung gears.

At the other extreme, a new design may have no experimental information available and must work from specified accuracies to guess the final T.E. This is extremely uncertain and modelling can then only be used to compare design on the assumption that errors will be similar regardless of design.

Usually there will be available information on helix, profile, and pitch accuracies obtained on similar gears together with an idea of how well alignments are held in practice from bedding checks. This information can then be used in a finite element approach to determine T.E.; such computations are large and are subject to commercial security so that it is sometimes necessary to use more primitive methods. At a particular instance the possible position of the contact lines is known; on any one tooth there is a single line of contact, as

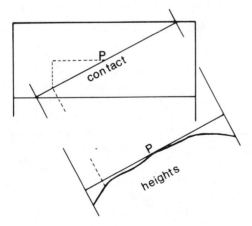

Figure 10.6.   Deduction of tooth surface "heights" relative to an involute through the pitch point P from helix and profile information.

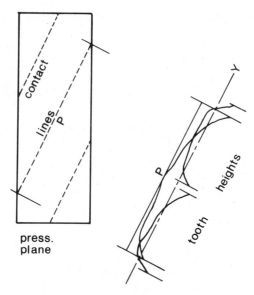

Figure 10.7.   Deduction of load distribution from contact lines and tooth "heights" relative to an involute through P. The area between a tooth "height" line and the line Y, multiplied by the tooth stiffness gives the instantaneous load taken by the tooth.

sketched in Figure 10.6, in this case shown running through the pitch point. Using the measured profile and helix information the "heights" of the other points on the line of contact above or below the involute can be deduced relative to the pitch point as sketched. Corresponding curves can be drawn for the contacts on the preceding tooth and the following tooth by working from the pitch point on the center tooth and using a pitch error to determine the "height" of the pitch point on the neighboring tooth then working along the helix (with errors) and up or down the profile (with errors) to determine relative heights. The curves obtained (3 in this example) make up the total possible length of the line of contact as shown in Figure 10.7. The actual length of line of contact is found by lowering the line Y until the area underneath the curves, multiplied by the tooth stiffness per unit length, equals the applied load. The distance of the line Y down from the pitch point represents the elastic deflection at the pitch point.

The mesh can now be moved through a quarter of a tooth etc., and the estimate repeated to find the new deflection of Y, relative to the original pitch point P and the variation of deflection of Y relative to P gives the T.E. This has assumed that the "other" gear is perfect but in practice its errors can be added in to give the overall T.E. The method is laborious and does not make allowance

Figure 10.8. Effect of deflection on T.E. At C both pairs of teeth are in contact (equally if under load).

for buttressing effects on local tooth stiffness but gives an idea of how much the elastic deflections of the teeth smooth out the manufacturing errors.

Another possibility is that T.E. information is available from inspection of the gear; this will be the T.E. at no-load. In the case of helical gears it is not possible to deduce full load T.E. from the no-load T.E. since it is not known whether the measured errors are due to local variations which tend to be averaged out by the elastic deflections or due to displacement of the whole tooth where there are no averaging effects. As a general rule the full-load T.E. can be expected to be better than the no-load T.E.; it is easy to postulate cases where the opposite will be true particularly where tilting of the gear axis occurs. A reasonable compromise is simply to assume that the loaded T.E. will be the same as the unloaded T.E.

Spur gears are a particular case where the effect of loading can be deduced since there is little interaction between helix and profile effects. Figure 10.8 shows a measured T.E. (full line) and also shows the part of the T.E. that may be deduced from knowledge of the gear profiles (shown dashed). Making the assumption that contact is across the full facewidth and neglecting variations in mesh stiffness we can state that under load the sum of the deflections of two teeth multiplied by tooth mesh stiffness must equal the applied load. As the sum of the two individual mesh deflections stays constant (or the one deflection in the area of single contact) it is easy to plot the elastic deflected position. A quick check that can be made very easily is to compare the elastic deflection during single contact (P/Kt), with the elastic deflection at the crossover point C which is where the two teeth are simultaneously in contact at no load and take $P/2$ each at full load. The difference in the elastic deflections is $P/2Kt$ and this should be equal to the measured drop in the T.E. at the changeover point C.

IV. LARGE SYSTEMS

We have so far considered systems which could be idealized as inertias connected by springs but this "lumped" approach can no longer be used when "large" systems are encountered. The term "large" is not definable exactly but refers in practice to systems where the wavelength associated with stress waves at the frequencies of interest is no longer large compared with body dimensions; it might be clearer to call such systems "short wavelength" systems. It is important to realize that a given design may behave as a lumped mass system for frequencies of once per revolution but as a large system for third harmonic of tooth frequency.

Direct compression waves in steel travel at 5000 m/s so that at 1250 Hz the wavelength is 4 m and only very large (physically) structures behave as "large" for vibration. In bending, the wave speeds are very much lower (2) and vary with frequency; at 1250 Hz in 10 mm steel sheet the wavelength is 280 mm, the same as the sound wavelength in air at that frequency.

With this type of system, non-linearity is usually not important but it is not realistic to model in a "lumped" form and so different techniques must be used. The basis of the approach is the use of "receptances" or responses of individual parts of the system; these can be measured experimentally or predicted theoretically and the system is then built up from the sub-units. This method is particularly useful for predictions of the effects of using different internal shaft and gear designs in a given gearcase which can be measured experimentally.

The response or receptance β_{ij} is the vibration generated at position j due to unit vibrating force at position i; directions must be specified and the response varies with frequency. In practical cases the response has damping to give a quadrature term as well as the in-phase response. The information can be held in a computer as an in-phase and quadrature response for a series of frequencies covering the operating range. Mobility is a particular case where the velocity per unit force is used but for large systems it is preferable to use acceleration per unit force and the inverse of the figure obtained gives the effective mass of the system at that frequency.

Figure 10.9 shows an idealized version of a 2-dimensional gear mesh with the gears A and B supported on a "large" structure S which is itself on antivibration mounts at positions 6 and 8. At the bearings, there will be horizontal and vertical components of force F but at the gear mesh the direction of force is assumed solely along the line of thrust. The response at any position is obtained by summing the responses obtained from all the forces acting on the sub-unit so if R is the response (acceleration) at the point and subscripts v and h correspond to vertical and horizontal then:

$$R_{1A} = F_1 (\beta_{11})_A + F_2 (\beta_{12})_A + F_3 (\beta_{13})_A$$
$$R_{1B} = F_1 (\beta_{11})_B + F_4 (\beta_{14})_B + F_5 (\beta_{15})_B$$

and $R_{1A} + R_{1B}$ = The transmission error. At the bearings, the shaft deflections are again given by summing responses so vertically for 2:

$$R_{2A} = F_2 (\beta_{22})_A + F_1 (\beta_{12})_A + F_3 (\beta_{23})_A$$

and the corresponding structure deflection which is equal is given by:

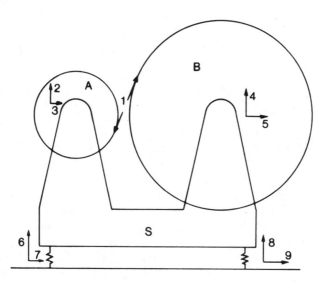

Figure 10.9.   Idealized 2-D single mesh. Gears A and B are supported by a structure S.

$$R_{2s} = F_2(\beta_{22})_s + F_3(\beta_{23})_s + F_6(\beta_{26})_s + F_7(\beta_{27})_s$$
$$+ F_8(\beta_{28})_s + F_9(\beta_{29})_s + F_5(\beta_{25})_s$$

In total, 14 such equations are required in this particular case and involve the measurement or prediction of 34 receptances; fortunately $\beta_{12} = \beta_{21}$ in a linear system so only half the number of cross-receptances need be measured. Certain cross-receptances such as $(\beta_{23})_A$ should be zero in a symmetrical system.

The computational effort is high even in this simple case and so when possible it is advisable to test in as large sub-assemblies as possible. In this example treating B and S together as a sub-system greatly reduces complexity and if further the Forces F_6 and F_8 to the foundation are not required only 6 equations need be used and 12 receptances measured or predicted. The effects of design changes to A, e.g., altering shaft stiffness can then be investigated since the changes only affect the receptances of sub-system A.

A simple example of adding a sub-system occurs when a "point" mass is added to a structure, or local mass is increased. The response of a point on a structure at 300 Hz is measured as 2 micron amplitude for a force of 100 N and lags $30°$ behind the force; what will be the response if an additional mass of 10 kg is added?

The common feature to both structure and added mass is the amplitude of vibration so assume unit amplitude of vibration (at $0°$ phase). The original structure then requires 5×10^7 N leading displacement by $30°$ and the mass requires

$10 \times (2\pi \times 300)^2$ N at $180°$ phase, i.e., 3.55×10^7 N. Adding these forces gives 2.62×10^7 N at $72.7°$ leading so the system response will be 3.9×10^{-8} m/N lagging $72.7°$.

V. THE STATISTICAL APPROACH

At high frequencies in "large" systems the resonances occur so frequently that there may be a very large number of resonances within, say, a 10% frequency band. Conditions then change so rapidly within 1% frequency range that the techniques described above become extremely laborious and do not give a clear picture of the response of the system. The best approach with many interacting modes close together is to work with system energy over a band of frequencies.

Instead of an input force producing output displacement we consider the system as a device into which we inject energy and which then either dissipates the energy internally or transmits it to its supports. The characteristics of the system are described by averaging the effects of many modes over a range of frequencies and the loss of energy in vibrating relative to the stored energy in the system is of major importance. Reference (3) describes the methods used and the approach adopted; this method is not suitable when resonances are easily separated.

VI. SCALE MODELS

The use of scale models tested using electromagnetic vibrators is well established in branches of engineering such as machine tool structures and the scaling rules are well known. The group $w^2 L^2 \rho/E$ is non-dimensional so that corresponding frequencies for model and full size are given by

$$\frac{w_1}{w_2} = \frac{L_2}{L_1} \left(\frac{\rho_2 E_1}{\rho_1 E_2} \right)^{1/2}$$

(or with shear modulus instead of direct modulus).

The stiffness of a model is proportional to the modulus times the size, i.e., to $E \times L$ and quarter size steel model will have a quarter of the local stiffness of the full size gearbox.

Problems with models arise at joints and bearings since it is difficult to get these representative. The use of models has declined as finite element techniques have become more economic though models are possibly slightly more reliable and have the added advantage in transparent plastics of giving a very clear understanding of the stress paths in a gearcase and of helping in piping layouts etc.

It is tempting to make a working scale model of a gearbox so that it may be run and can then give information directly about alteration of alignments etc. Reducing size has the effect of raising natural frequencies so it is then necessary to run the gearbox faster to get the correct relationship between forcing and natural frequencies. A quarter size model of a large gearbox would have 2mm module gears instead of 8mm module and would transmit one sixty-fourth of the torque but as it would run four times faster would still need a sixteenth of the power. The major problem with such scaling is that accuracies do not scale and much the same accuracies are attainable regardless of size; a full size box might have accuracies of 3 to 4 microns and it would be unreasonable to expect better than 1 micron in a model.

Ideas of scaling may be useful to predict dynamic performance in a range of similar gearboxes. If a new box, say 30% larger on dimensions than an existing box is being proposed then a good idea of its vibration response may be obtained by running the existing box 30% above speed to get frequency ratios correct.

REFERENCES

1. Timoshenko S., Young D. H., and Weaver W., Vibration problems in engineering, Wiley 1974 (4th. Ed.)
2. Drago R. J. How to design quiet transmissions. Machine Design. Nov. 20th 1980, pp. 114–120.
3. Cremer L., Heckl M., and Ungar E., Structure borne sound, Springer-Verlag, Berlin 1973.

Chapter 11

DEVELOPMENT PRIORITIES

I. CASE DEVELOPMENT

When presented with a noisy gearbox the first step is usually to take a pickup over the casing. This is simply because the casing is easily accessible and can be checked quickly and easily to see if there are obvious defects such as resonating panels. The measurements have been mentioned in Chapter 9 and the main objective is to check whether either the gearbox is extremely rigid and is moving as a nondistorting block or, alternatively, if it is well above a natural frequency so that the panel itself is rigid but is moving with less amplitude than the main gearcase. A criterion which works well in most cases is that ideally no point on the case should be moving more than the bearing housings through which the gear vibrations are transmitting.

In addition to running the gearbox and measuring amplitudes it is sometimes worthwhile exciting with a vibrator at a bearing housing (in the expected direction of forcing) and measuring amplitudes round the case. This will give the mode shapes in a pure form unlike the irregular excitation from gears and will also allow estimation of the local impedance of the gearbox to determine the observed effective mass.

The theoretical limits of performance for a gearcase can be estimated simply if it is mounted on soft isolators since reduction of vibration is limited by the actual mass of the gearcase. It is relatively easy to estimate the effective mass that should be measured if a gearcase is rigid and the observed mass can be expressed as a fraction of this. If a gearbox is idealized as a cube of mass M, side 2a and radius of gyration k, then vertical excitation in the center of one side as in Figure 11.1 should "see" a mass of $Mk^2/(k^2+a^2)$ while vertical excitation at a corner should see a mass of $Mk^2/(k^2+2a^2)$.

When the gearcase is rigidly attached to a large unit, then no simple criterion is available. At high frequencies the restraint attachment can probably be ignored and effective masses should be as above, but at low frequencies the impedance is dominated by the rigidity of the main structure.

At a few specific frequencies effective masses may be greater than actual masses suggesting that there is a resonanace somewhere in the system acting as a

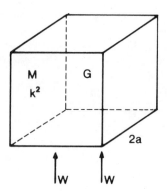

Figure 11.1.   Effective mass of gearbox. If a rigid body is excited the measured movement at the force is due to lateral and torsional effects so the apparent mass is reduced by a factor of $(k^2 + a^2)/a^2$ or $(k^2 + 2a^2)/a^2$ at a corner.

vibration absorber; these should be checked since case resonances will radiate well and internal resonances give high forcing.

II. T.E. MEASUREMENT

Whenever possible T.E. measurement under load is very helpful because it gives an immediate clue as to whether manufacturing or design is primarily responsible for a noisy box. When this is not possible, one of the approaches described in Chapter 10, Section III should be used to estimate the T.E. of the drive.

The T.E. is liable to vary greatly even for successive gears from an automated line so it is as well to check several gears whenever possible. Variations of noise levels from mass-produced gearboxes or back axles is usually due to variation of T.E. rather than variation of system dynamic response. Measurement of several meshes will show whether there are consistent errors and should give strong clues to any manufacturing problems.

The difficult question then arises as to permissible levels of T.E. from manufacturing. Economically for mass production, if it is possible, it is nearly always cheaper to achieve noise objectives by improving casing or internal design, since the initial development cost is spread over many units whereas on a 1-off it is cheaper to improve gear quality, if it is possible. There are in practice limits beyond which it does not seem possible to go, almost regardless of the time and money that may be spent. As a rough rule it is possible to hold pitch, profile, and helix each to within a 2 micron band if cost is no object. The resulting T.E. can be within 2 microns at once per tooth frequency if extreme care is taken in assembly and the alignment is optimized by testing T.E. under load. Then, for this loading level only, it is unreasonable to expect lower excitation.

In normal engineering there must be an economic balance between gear manu-facturing and gearcase costs so it is advisable to have rules of thumb as to what is reasonable in a particular industry based on experience. Large facewidths with high helix angles give good averaging effects along the helix so it is reasonable to expect once-per-tooth errors of the order of 5 microns and similar pitching errors from ground gears regardless of load levels. On narrow gears end effects become more important and so, although pitching errors should be within 5 microns, the once-per-tooth errors are liable to reach 10 microns (peak to peak).

Spur gears when ground will have similar pitching errors but the once-per-tooth errors increase greatly away from design load and can easily reach 20 microns on a heavily loaded gear. At design load the once-per-tooth errors are likely at best to be about 5 microns.

Unground gears in the "soft" state after hobbing can be accurate in pitch and helix but are less accurate on profile so that once again the observed T.E. on a helical gear will be very dependent on alignment but on a spur gear may be up to 20 microns.

It is possible to reduce measured T.E. in a design by going to high helix angles which, by increasing the number of axial pitches within the facewidth, give better averaging and, hence, decrease T.E. The trouble with this approach is that the axial forces increase with helix angle except with double helical gears or gears with thrust cones and so the excitation in the axial direction becomes more important. Unfortunately, most gearboxes are more sensitive to axial vibration than to radial vibration so the effect is large.

III. INTERNAL IMPROVEMENTS

After investigating case vibration and checking T.E. the next stage is to see whether the vibration reaching the bearings can be reduced. Resonances within the working range are an obvious fault and should be attacked but it is often found that there are no significant resonances occurring within the speed-range.

It is necessary to have a model of the behavior of the gears inside the box to be able to predict the effects of changes. In complex cases full analyses are required but the simple system model discussed in Chapter 8 is often sufficient. It is rare in practice that mass can be altered due to lack of room but relatively small changes in shaft lengths or diameters will alter stiffness greatly since cube or fourth power laws are involved.

Well below resonance reduction of gear support stiffness will reduce trans-mitted vibration since a given exciting deflection will generate less force in a soft spring than in a stiff spring. Similarly, well above resonance, the amplitude of vibration of the gear body is fixed by its inertia at a given frequency and reduc-tion of support stiffness will reduce transmitted forces so that it would appear, at first sight, always advantageous to reduce stiffnesses.

Problems arise because though support stiffness reduction helps below the first resonance and above the highest resonance it can make matters worse between resonances and there is danger of bringing resonances near excitation frequencies. Only modelling of the system will allow prediction. Some knowledge of the sensitivity of the installation to particular frequencies is required since it often happens that vibration at tooth frequency can be reduced at the expense of increasing vibration at other frequencies such as once per revolution or third harmonic of tooth frequency.

The other limitation on reduction of gear support stiffness arises from the static deflections of the gears under the steady load. Taking simple figures of an effective mass of 2.5 kg, subject to a tooth frequency excitation to 700 Hz, what will be the static gear deflection under a tooth load of 5000 N (0.5T) if the support stiffness is chosen to give resonance a) above or b) below tooth frequency?

In case a), to give a safe separation from the exciting frequency the support stiffness should keep the resonance above 1000 Hz. This gives a stiffness of not less than 10^8N/m and a static deflection of 50 microns (.002″). Case b) requires a resonant frequency below 500 Hz so that the corresponding stiffness is 2.5×10^7 N/m giving a deflection of 200 microns. The amplitude of vibration of the gear mass is roughly the same in both cases so that case b) will transmit roughly a quarter of the vibration amplitude at the penalty of a static deflection increased by four. This very simple approach is complicated in practice by the presence of several degrees of freedom and non-rigid bearing supports but the conflict between vibration levels and static deflections exists.

Permissible static deflections in a system are open to much debate; the figures obtained by estimates are initially the deflections in the direction of the line of thrust and so must be multiplied by sin 20° (or rather the transverse pressure angle) to get the movement apart of centers. The involute form was chosen for its tolerance of center distance so that a gear drive should be able to tolerate center distance movements of perhaps 10% of the normal module without distress since only some 5% reduction of contact length would be involved. Most gear designers would however, be unhappy at the thought of allowing 0.6 mm center movement apart, i.e., 1.7 mm of centerline deflection on a 6 mm module gear. A more realistic limit might be 2.5% of the normal module since this would have negligible effect on the mesh; this corresponds to a deflection of just over 0.4 mm for a 6mm module gear.

In double helical gears an increase in helix angle will affect mesh stiffness in the direction of the pressure plane. The length of line of contact is proportional to contact ratio times sec of the base helix angle; for movement in the transverse plane tooth deflections are reduced by a factor $\cos\alpha_b$ and the corresponding normal forces by another $\cos\alpha_b$ so the overall effect is to affect stiffness by a factor of $r_c \cos\alpha_b$. Even helix angles as high as 45° will only reduce mesh stiffness by about a factor of 2 which usually has little effect on natural frequencies.

The other form of stiffness alteration which is occasionally possible in lightly loaded gears is the use of plastics such as nylon or acetal. As mentioned in Chapter 8, the large reduction in mesh stiffness gives very much quieter running despite some increase in T.E.; the frequency of the higher resonances of the system are greatly reduced which usually helps noise levels and the excitation is less powerful at all frequencies above the first resonance of the system.

It should be emphasized that stiffness reduction must not be carried out if it brings resonances near powerful forcing frequencies and so it is an approach that can only be used if the vibration characteristics and excitation are known or predictable.

IV. ALTERATION OF EXCITATION FREQUENCIES

An alternative approach to separation of natural frequencies, i.e., resonances and excitation frequencies is to attempt to move the excitation frequencies to a less troublesome area. Speed is fixed so that once per revolution frequency is fixed but it is rare that once-per-revolution is important in its own right; it usually only appears as a modulation of tooth frequency or harmonics. The only variable available is alteration of the number of teeth on gears to shift tooth frequency.

Reduction of tooth frequency is rarely considered but can pay large dividends if the tooth frequency is relatively low, say below 500 Hz and the system is operating below its main resources. Under these conditions, with fixed amplitude of vibration excitation the output amplitude and the sound level will be proportional to frequency squared so that a 30% frequency reduction at constant load will give 6dB sound pressure reduction. There is, however, a bonus since the ear's sensitivity reduces with frequency below 500 Hz so there is a further gain of 3dB in the A weighted sound level.

The corresponding penalty lies in increased metal removal costs since the module has increased and in some decrease in accuracy; this is difficult to quantify but a rough rule of thumb for industrial gears is that the errors are proportional to the square root of the size. Large gears for marine use do not seem to obey this law and manage to achieve nearly the same accuracy for 5 m diameter as for 100 mm diameter. The increased tooth size does not affect contact stressing and gives reduced root stresses but gives higher sliding velocities and so greater tendency to scuff.

At high frequencies increase of tooth frequency by reduction of module should be investigated. In this regime the assumption is that excitation is above the natural frequencies of the system and so the displacement amplitude will be inversely proportional to frequency squared. Emitted sound will then be inversely proportional to frequency squared on the assumption that sound depends on velocity squared.

Reduction of module will ease scuffing and leave contact stresses unchanged but will increase root stresses and so must be approached with care; in many designs there is a very large margin of safety on root stresses, allowing reduction of tooth size.

In a two-stage reduction gearbox the choice of ratios from the noise aspects is very complex. Second-stage tooth frequency is relatively low if, for example, the output speed is 180 rpm, keeping the output wheel number of teeth down to 60 would keep mesh frequency below 200 Hz but would restrict the final drive ratio to about 2.5 to 1. This in turn would require a large first stage ratio, 10 to 1 if the overall ratio was 25 to 1; the mesh frequency of the first stage, using 24 teeth would be 1800 Hz. This would be a more costly drive than using two stages of 5 to 1 but would give appreciably lower noise levels from the second mesh whilst not affecting first stage noise.

Use of an epicyclic gear instead of a conventional drive for speed reduction gives a slight lowering of tooth frequencies due to the epicyclic effect but the change is small, typically only 25% and is completely swamped by the large noise reductions due to the inherent force balance in the system which allows input and output shafts to be well isolated and allows extremely effective isolation of the annulus without disturbing gear alignments.

V. DAMPING

When resonances or "ringing" sounds are encountered the question of damping and its possible effectiveness arises. Damping is extremely unpredictable and rather temperamental but it is worthwhile looking at the various mechanisms of damping to see if a particular method might work.

Steel is an extremely elastic material with internal damping which is so low that it may be ignored for all practical purposes. Cast iron is much better from the damping aspects and much is made of its properties relative to steel but its damping is still much too low to be sufficient for resonance control; the dynamic amplification factor (Q) associated with cast iron is of the order of 100 whereas we want less than 5. High damping can be obtained with some Manganese-Copper alloys but these are not practicable because they need to work at high stresses and are not economic.

Rolling bearings give negligible damping for all frequencies and as far as noise is concerned their only asset is that they provide a gap in the transmission paths for high frequency pressure pulses. Hydrodynamic bearings give very much more damping and have a large effect on one-per-revolution frequency but at high frequencies such as tooth frequency the bearing squeeze-film stiffness has become so high that the oil film stays rigid while the metal round it deflects.

Micro-slippage at joints with the associated friction losses give what is probably the major form of energy loss in isolated gearboxes but it is difficult to predict. It is sometimes found by experience that tightening up bolted joints improves damping but just as often a slight loosening of bolt tension may improve the response. Attempts have been made to improve joints by inserting layers of damping material but the end results have not been satisfactory, probably because allowing sufficient movement to absorb low levels of vibration energy would mean high levels of static deflection. This is a fundamental problem when static forces are a thousand times dynamic forces. Welding can provide damping at joints if the welding is incomplete or there are shrinkage cracks present; stressing is often not a problem in gearboxes since the main requirement is rigidity, not strength, and welds which do not have full penetration should be considered as they are not only very much cheaper but provide damping.

When a gearbox is not isolated but is attached to another, usually larger, structure the main mechanism of damping appears to be a radiation of energy from the gearbox. Stress waves from the vibration carry energy away from the gearbox and although there may be no obvious absorption mechanisms in the main structure, the energy does not return. In this case alterations to the gearbox will not affect damping; the better the mechanical coupling between gearcase and structure the better will be the damping observed in the gearbox but the energy will cause trouble elsewhere.

Increasing damping in an existing gearbox or part of a gear drive is usually done by adding high friction joints, viscoelastic layers on panels, or vibration absorbers. The most common form of high friction joint is to shrink or to bolt a wheel rim onto its supporting webs instead of welding; this has the effect of damping many of the internal modes of vibration of the complete wheel. Splitting a gearcase can have the same effect.

Viscoelastic layers are much used to prevent vibration on car body panels but are less effective on gearcases. For a damping layer to work well the dynamic stiffness in bending of the viscoelastic layer should be not less than about 5% of the stiffness of the panel to which it is attached. This is easily achieved on a car body panel where the added layer can easily be 4 times the thickness of the metal but on a rigid gearcase with a wall thickness of 12 mm the corresponding viscoelastic layer thickness of 50 mm would be ridiculous. The effectiveness of the layer probably owes much to pure mass effects reducing frequencies as well as its energy absorption properties.

Vibration absorbers are little used in gear work though they are widely used for machine tool vibration control. Their theory and effectiveness is well discussed in Den Hartog (1); assuming an absorber mass of 10% of the main mass can be fitted, a damped absorber will give a dynamic amplification (Q) of 20 whereas a tuned damped absorber will give a Q of 4. It is rare for gear systems to have very low damping and so only the tuned damped absorber should be considered; a Q of higher than 10 on a single troublesome resonance is required for an absorber to be worthwhile. Damping and tuning is best achieved by using a nitrile rubber support which can be distorted to allow frequency tuning to the optimum frequency. This method is useful for very sophisticated applications but is expensive because it requires time and skill to set it up and the performance is liable to drift with time and temperature.

VI. ISOLATION

In the majority of cases, noise from a gear mesh causes trouble not by direct case-radiated sound but by transmitting vibration through the gearbox supports and shafts to the remainder of the structure. The vibration isolation of the gearbox or, in the case of a vehicle, the gearbox and engine block becomes extremely important.

Problems in isolation design can be split into two main families; one batch of problems is associated with selection of layout geometry and isolator stiffness so that the six possible degrees of freedom are decoupled, i.e., do not interact. A prime example of the need to decouple occurs in a vehicle engine and gearbox assembly since there will be very large torsional vibrations occurring at idling due to the engine characteristics; it is vital that these large torsional vibrations are not coupled to lateral movements or the resulting vertical or sideways vibrations of the engine and vehicle would be very uncomfortable for passengers. A quick, practical test for interaction in a system is to depress the center of gravity and observe whether rotation occurs or vice versa. Full analysis of complex layouts is not of relevance for gear induced vibrations since the effects of coupling are not important but should be carried out for low frequency vibrations.

The dominant problem in noise isolation is that isolators which are designed primarily for low frequency effectiveness will transmit vibration well at high frequencies. A soft spring will begin to "surge" and behave as a distributed mass system as the frequency rises and so its characteristic will not be as shown by the theoretical full line in Figure 11.2, but will oscillate about a limiting value perhaps 20dB down on the static value. The presence of extra mass in the middle of the spring, as occurs when steel plates are used to laminate rubber springs, will reduce surge frequencies to very low levels. We can estimate natural frequencies when the geometry is simple and this gives a good idea of the permissible frequency ranges for operation.

A spring which is resonating, whether longitudinally or laterally, will have a deflected wave shape which is a half sine wave (1). With stiffness per unit length S (N/m/m) and a mass of m (kg/m) permit length, the wave speed will be $(S/m)^{1/2}$ and an unsupported length of ℓ will be half a wavelength, i.e., the wave speed divided by twice the frequency. For rubber in shear the value of G is of

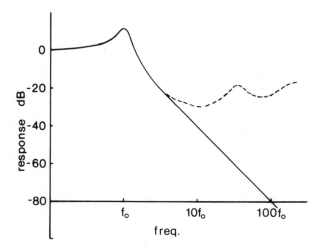

Figure 11.2. Effect of spring surge on isolation effectiveness. The full line is the theoretical single degree of freedom response but the observed response is given by the dashed line.

the order of 10^6 N/m^2 (150 lb./sq. in.) for a medium hard rubber so the value of s per unit cross-section is 10^6 N/m/m^2/m, and the density is 1300 kg/m^3 so the wave speed is about 27 m/s. This gives a surge frequency of about 270 Hz for a wavelength of 100 mm, i.e., a pad thickness of 50 mm. Steel coil springs have low mass per unit length but corresponding lower stiffnesses and so can give comparable wave speeds.

Surge problems are inherent in soft springs and cannot be removed by change of design or material so where both high frequency and low frequency vibrations are involved it is necessary to have two stages of isolation. This is shown in idealized form in Figure 11.3; an additional mass is required and will resonate at an intermediate frequency which must be kept well clear of excitation frequencies. Occasionally an existing part of the machinery such as a sub-plate can be used as the intermediate mass. Choice of whether the "soft" or "hard" spring is used beneath the sub-plate can depend on the frequencies of the other excitations on the sub-plate but it is usual to have the softer springs under the sub-plate since the greater mass then allows harder springs and, hence, less deflection under load. Selection of a natural frequency for the resonance of the intermediate mass depends on what other excitations may be encountered but is typically at 3 to 4 times revolution frequency and, hence, still a factor of 5 down on tooth frequency. In contrast the main isolation will ideally be at about a third of revolution frequency. A two stage gearbox becomes complicated as there are 3 rotation frequencies and two mesh frequencies to be avoided.

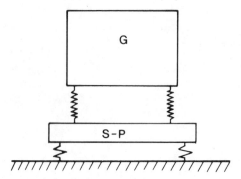

Figure 11.3. Use of sub-plate to give 2 stage isolation. G is the gearbox and SP the sub-plate which is also isolated. Either set of springs may be of low stiffness.

In Section IV the gearbox mentioned had tooth frequencies of 1800 Hz and 200 Hz and shaft frequencies of 75 Hz, 7.5 Hz, and 3 Hz. It is unrealistic to attempt to isolate the two lowest shaft frequencies, and these frequencies are so low that the noise is not important so the main isolation frequency chosen might be about 25 Hz for vertical vibration. The soft isolation chosen for this would not work well at 1800 Hz so an extra stage would be required with a difficult choice of natural frequency. The likelihood is that the 200 Hz vibration is the more powerful vibration but there is insufficient room between 75 Hz and 200 Hz to fit an additional isolation stage with confidence, particularly if speed varies. An intermediate plate with frequency set about 300 Hz would give good attenuation of primary mesh frequency but not affect 200 Hz, so some development of the main isolators to avoid surge at 200 Hz might be necessary.

Direct air borne sound can best be isolated by putting a sealed case round machinery; small gaps let out a large amount of sound and so must be sealed but any mechanical connection between gearbox and sound case will transmit noise effectively. Oil pipes, though flexible, are very good at transmitting vibration. Acoustic hoods such as those used for telephones and typewriters are surprisingly effective for high frequency sound but of little use for low frequency sound.

VII. CANCELLATION

When all else has been tried and is insufficient the use of sound cancellation can be considered in cases where only a small region needs to be kept quiet. Active cancellation is a subject which is still developing and requires sophisticated equipment and skilled labor and so is expensive. There are several different approaches (2).

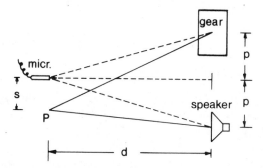

Figure 11.4. Effect of sideways movement on sound cancellation. Perfect cancellation is assumed to exist at the detection microphone and the sound level is required a distance S to the side.

One basic principle is shown diagrammatically in Figure 11.4; a loudspeaker is placed near the gearbox and radiates a sound which is the same as that of the gearbox but in antiphase so that the two sounds cancel. If the sound repeats exactly each cycle and there is no interfering background sound, then at the microphone it is theoretically possible to get complete elimination. The practicalities of variations in the sound output and interference put a limit on the reduction that can be obtained at the microphone but rather more important is the difference in position between microphone and observer.

Assuming that the effective center of the gearbox and the loudspeaker are the same distance from the microphone, the two sound waves will be in antiphase and will have the same amplitude at the microphone. At a point P to the side of the microphone the path lengths will be different so that the waves are not exactly in antiphase. The combined result will be a wave whose amplitude is that of the original sound multiplied by twice the sine of half the phase difference (cos A - cos B = 2 sin (A + B)/2 sin (B - A)/2). A phase difference of less than 36 degrees is required to give an attenuation of 4dB on the original signal so that at most we may have a path difference of one tenth of a wavelength. At 200 Hz with a sound speed of 300 m/s this allows 0.15 m path length difference. This path difference must then be related to the geometry of the system to find the permissible sideways movement between P and the microphone. Using Pythagoras, expanding with the binomial and approximating gives the path difference as $((p+s)^2+d^2)^{1/2} - ((p-s)^2)^{1/2}$ which reduces to 2 ps/d which is the sideways movement s multiplied by the tangent of the angle subtended by the gearbox and loudspeaker. In a particular case if the separation is 1 m and the observer is 5 m away then a path difference of up to 0.15 m allows sideways movement of 0.75 meters, which gives a large area of reduced noise. The total power emitted is however doubled and there will be areas where the sound level is 6dB increased.

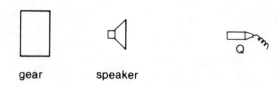

gear   speaker

Figure 11.5.   Effect of distance on in-line cancellation. Perfect cancellation at Q will give incomplete cancellation elsewhere along the line due to inverse square law effects.

Movement nearer or further away from the gearbox will give relatively little change in observed sound level since this is a symmetrical geometry. At a position such as Q in Figure 11.5, the system is relatively insensitive to sideways movement but is more sensitive to movement nearer the sources since the loudspeaker power level must be lower than that of the gearbox to give cancellation and the inverse square law will mean that the contribution from the loudspeaker drops faster with distance than the contribution from the gearbox. The observed power level can be deduced directly from the inverse power law; taking the original noise level as 84dB 2 m away and interposing perfect cancellation 1 m in front of the gearbox requires a loudspeaker power 6dB less than the gearbox. Even large movements to, say, 2 meters further away will give levels of 78dB from the gearbox and 74.5B from the loudspeaker, cancelling to give in theory a sound level of 67dB. The performance in practice will be very much poorer than this due to reflections and irregularity but this represents a target performance under perfect conditions.

One loudspeaker, suitably programmed, can in theory reduce sound in one restricted area and it can be inferred that use of many loudspeakers will allow sound reduction in many areas but the analysis and computing problems are large so that economics rule out this approach for gearbox work and total sound power rises for normal cases.

Vibration cancellation is in theory very much easier since there are a finite number of paths for the vibration. In a simple case a gearbox might be supported on 4 antivibration mounts which were stiff vertically but soft sideways so that cancelling would require 4 vibrators mounted vertically at the support points. Interaction between the vibrators would require complex computation as with sound level fields but as only 4 detectors and 4 vibrators are required to give complete isolation the problem is theoretically soluble. In practice the necessary vibrators are rather heavy and somewhat delicate; they must be supported off the gearbox on relatively soft supports and so are extremely vulnerable to damage and will not work effectively at very low frequencies.

REFERENCES

1. Den Hartog J. P., Mechanical Vibrations, McGraw-Hill.
2. Ross C. F., Active control of sound, Ph.D. Thesis, Cambridge University Engineering Department, 1980.

Chapter 12

STANDARDS

I. NATIONAL STANDARDS

The principal gear standards in use are the AGMA 170.01, the British BS436, the German DIN 3990, the corresponding I.S.O./DP 6336, and the proposed new British standard. Other publications such as AGMA 299.01 are advisory or are concerned with methods of noise measurement. These standards are not primarily concerned with noise and vibration but with permissible loading on gears. Many of the questions that should be asked when rating gears are however the same questions that should be asked when investigating noise so it is worthwhile looking more closely at the standards.

The "old British" approach to gear rating can best be described as the "averaged experience" method, whereby gear manufacturers and users collaborate to provide extremely empirical rules of thumb based on operating experience. Permissible loads are specified for "typical" manufacturing accuracies of a given class with "typical" loading cycles and corrections for speed, etc., follow a standard law regardless of system dynamics, methods of manufacture, etc.

In complete contrast DIN 3990 (and 1S0 6336 which is derived from it and the proposed new B.S. specification also derived from it) is firmly based on much research work and attempts to make exact calculations for what is going on from measured or estimated figures. This standard does not work on "typical" figures for, say, misalignment but uses estimated or actual figures to predict the load increases expected as in Chapter 7, Section III. It is a very much more complex standard than previous standards because every factor is investigated (in theory) exactly. It is no longer sufficient to specify "average commercial quality" of manufacture but precise details of helix, pitch, mounting, and profile errors are needed for the calculations. The resulting amount of work involved in designing a gear is very much greater than that necessary for the older methods and involves asking a large number of questions to which the answers are not available. Criticisms can be, and are, levelled against the DIN approach because although much of the detailed work is very good and thorough, some of the assumptions are very imprecise or naive; in particular the approach to system dynamics is very weak and should not be used.

The AGMA 170.01 standard is a compromise approach which has attempted rather more than the old standards but has fallen short of asking all the embarrassing questions about exact levels of gear accuracy and system dynamics so that it should still be classified as an "industrial averaged" approach. This leaves it essentially as an imprecise approach.

II. EFFECTS OF STANDARDS ON DESIGN

It would seem at first sight that it does not matter very much which standards are used for design as any standard will produce a gearbox that is unlikely to fail. There are likely to be severe repercussions on costs with the use of the more precise DIN approach since it highlights sensitive points in the design.

An initial design, assessed according to traditional standards, has a fixed load capacity which allows for "typical" errors and dynamics. There is no advantage in detailed investigation or control of any single factor since it will not affect the allowable rating of the gearset. In contrast the more detailed investigations associated with the DIN approach will emphasize particular points which control the load carrying capacity. This detailed analysis is typical of design and development in airplanes because the economic aspects of excessive weight justify the high initial development costs.

If we take a simple example of a gear loaded to 170 N/mm of facewidth the tooth deflection is about 12 microns and the typical control of alignment on assembly might be 25 microns so that overloads of 100% are involved. Control of alignment to 12 microns would reduce overload due to this cause to 50% and, hence, allow a 30% increase in rating. Similarly, operation with a T.E. of 12 microns peak-to-peak at once per tooth frequency will give 50% overloads if the system is above resonance but gives negligible effect if the system is well below its resonance; if the system is not checked dynamically there is a possible 50% increase in rating which may be wasted.

The overall effect of using a more precise standard is that much more time and effort must go into initial design work but that the resulting gearbox is likely to be more economical because it is smaller. This will, in time, give a competitive advantage to the manufacturers who can use their technical superiority to produce smaller gearboxes. Present tendencies are to automate gear design by using standard computer routines to carry out all the detailed work of calculation; this approach saves time but has the disadvantage that it is not immediately obvious from the end results what the important factors are in a given design. To overcome this, it is advisable to print out the intermediate factors such as the load distribution factors and the dynamic (internal) factor to see where possible improvements can be made. The alternative is to try a series of possible errors, support stiffnesses etc., to see how they influence the end result.

Despite the general tendency for the DIN approach to insist on exact calculations it occasionally lapses into "typical" figures as when it suggests that an additional application factor of 1.1 might be used for step-up gears, as in the AGMA specification, if no other information is available. This procedure should not be followed but account should be taken of the effects of driven inertia. If an electric motor is accelerating an equivalent inertia which is less than that of the motor itself, as normally applies in machine tools, the acceleration times are short and the torque through the gearbox on run-up will be less than the rated motor torque. A load with high equivalent inertia, whether due to gearing effects or actual inertia, will give a very much longer run-up time, some ten times longer, and the loads can be up to six times the rated motor torque so there is a double effect on permitted fatigue life. A given motor gearbox combination can have a life a factor of 100 different for the same steady state loading according to whether it has frequent slow accelerations of a large inertia or infrequent low inertia accelerations.

III. EFFECTS OF STANDARDS ON NOISE

Standards have no direct effect on noise but as they influence design and manufacturing control they will influence noise indirectly.

We have in Section II suggested that use of more precise standards in design, though initially costly, will highlight critical areas and, hence, tend to produce smaller, lighter and slightly more accurate gearboxes. Alignments and, hence, contact patterns should be better and there should not be resonances in the working range.

At first sight the effects of higher manufacturing accuracy and more even load distributions can only be an advantage from the noise point of view, since reduction in force variations helps noise as well as stressing. This is generally true and a better designed gearbox will normally be quieter than a competing gearbox. There is, however, one possible disadvantage in that a highly stressed gearbox will be smaller and have significantly less mass in its casing and in the gears themselves. From the vibration transmission aspect, reduction in mass is a disadvantage since a given force level will produce higher displacements. Stiffnesses are likely to be unaltered since shaft and bearing sizes are dictated by the loads which remain constant and so the natural frequencies of the internal system will move up; this is a disadvantage for larger gearboxes where the main forcing frequencies are above the natural frequencies of the system.

IV. ISO FACTORS

The DIN/ISO/proposed BS has a very comprehensive list of factors which in theory allow for all possibilities. Even though a designer has no intention of using this standard it is worthwhile having a list of the standard factors as a check list or reminder of what may influence a particular design. Using the ISO/DP 6336 the factors are, in the order in which they should be applied:

K_A Application factor. Allows for dynamics of load system to convert nominal external loads to dynamic actual loads.

K_V Dynamic factor. Allows for increase of loading due to internal dynamics increasing tooth loads.

K_β Load distribution factor. Accounts for variation of loading over facewidth of tooth due to helix errors, misalignments, gear distortion and case distortion (possibly reduced by $\gamma\beta$ the running-in allowance).

K_α Transverse load distribution factor. Accounts for uneven load distribution between successive teeth due to pitch and profile errors preventing equal load sharing. The limit for this occurs when all the load is taken by one tooth as with spur gears.

Once the actual maximum local tooth load has been determined using the above factors the individual stressing calculations can be carried out. As a rough guide K_V and K_β should be kept below 1.5 for efficient design; K_α should be kept down to below 1.2 or the stressing advantage of helical gears are being lost. A value of K_A above 1.5 suggests that detailed cumulative fatigue life estimates are needed.

The detailed contact stress calculations (Part 2) are tedious and are mainly concerned with corrections to be applied to the stress due to variations from the arbitrarily selected conditions which are pitch point contact, low helix angle, infinite life, viscosity at $50°C$ of 100 mm^2/s (100 centistokes), pitch line velocity of 10 m/s, peak to valley roughness of 3 microns (equivalent to 20 microinches C.L.A. roughly), and hard teeth.

Root strength calculations follow a similar pattern for standard conditions of $20°$ pressure angle, standard 1.25 m_n dedendum profile, low helix angle, 3×10^6 cycle life, low root surface roughness, 5 mm module gears.

Scuffing calculations (Part 4) are rather more open to argument since the underlying mechanisms are not fully understood and the behavior of oils, the corresponding coefficients of friction, and the local temperatures are all rather hypothetical. This area is so uncertain and the properties of extreme pressure oils on real surfaces are so difficult to measure that running experience under similar conditions is essential.

Although the ISO specification is better than its "averaged" predecessors in that it is more precise, it should be treated with caution and it is essential that intelligence is used in its application. It is far too easy for a design office to work on the assumption that gears will be fully corrected and expertly bedded at full load when the manufacturing side is making "commercial" quality gears.

Certain of the ISO factors such as roughness factor and lubrication factor for contact stresses are allowed to exceed 1.0 in the specification but common sense suggests that these be limited to 1.0 when designing. The ISO approach to stressing, like all other standards, ignores the effects of tip relief on stresses due to effective shortening of the length of the line of contact.

Chapter 13

GEAR FAILURES

I. INTRODUCTION

Gears are unusual in engineering in that they show a considerable number of different forms of damage, some of which are not important or can be cured easily. An inexperienced engineer may take out of service gears which would run happily for 20 years and yet ignore faults which will produce catastrophic failures rapidly. Damage may also give a clear idea of trouble in the system from causes other than the gears.

The common types of damage are described well in the Tribology Handbook by Neale (1) which gives photographs and detailed descriptions of the surfaces. Visual inspection is usually all that is possible and so much depends on the skill of the viewer.

Roughly in order of importance we may classify damage as follows but though pitting occurs most frequently, it is often not serious.

II. PITTING

Pitting is a fatigue effect which occurs due to higher Hertzian contact stresses than the surface can stand. The maximum shear stress occurs below the surface, starting a crack which breaks out and leaves a small, smooth bottomed crater. Pitting is usually arbitrarily divided into initial pitting and progressive pitting. The flank surface will start with a limited number of high spots; if the surface roughness is comparable with the oil film thickness there will be extra high pressure on these spots which will lead to rapid fatigue and the removal of the high spots.

This initial removal of the high spots occurs in most gears unless they are lightly loaded or have been made very accurately. Once the high spots have been removed and the load has been redistributed a correctly designed gear will not generate any more pits and can then function perfectly ad infinitum. Initial pitting usually occurs near, but not on the pitch line, in the region where the oil film is thin but there is still a sliding velocity.

Progressive pitting occurs when the load redistribution associated with initial pitting is not sufficient to reduce stresses below the fatigue limit and the new high spots are continuously removed by fresh craters. In extreme cases sufficient of the surface area is removed by pitting that the remainder of the surface fails extremely rapidly. Eventually the whole flank of the tooth can be destroyed.

The bounding line between initial pitting and progressive pitting is not an exact one but is sufficiently clear for most practical purposes that they can be regarded as separate. When a gearbox is opened for inspection, ideally after about 3 million load cycles for a long life box, limited area of pitting is usually satisfactory. If there is any doubt, another million cycles of load applied will allow checking of whether any further pitting has occurred; if not, the gear may be left for long periods.

Cure of progressive pitting depends on the cause. The most usual cause is insufficient surface hardness and fresh gears are needed with better case control. Poor load distribution may contribute to pitting which will indicate what alterations are required in alignment or corrections. If the cause lies in basic design, usually only increase in diameter or facewidth will give a cure; it is, however, nearly always more economic to improve materials if possible than to alter a design. Sometimes overloads on frequent starting can be responsible for damage and the easiest solution is then a "softer" start.

A particular form of pitting is spalling which gives breakaway of relatively large bits of hardened case. It can be due to too abrupt a transition between a hard case and the relatively soft material underneath or may be due to local metallurgical defects. Stress concentrations at tooth ends can also produce this effect.

III. SCUFFING

Scuffing occurs when the oil film breaks down to allow metal to metal contact; this gives local welding followed by a tearing of the surface as the welds are sheared.

There are two mechanisms which give oil film breakdown. Failure of the oil film at very low speed is caused by insufficient hydrodynamic bearing action to keep the surfaces apart. Little heat is generated since sliding speeds are low and damage is most likely to occur at minimum film thickness, i.e., near the pitch line. This form of scuffing is not common and is known as cold scuffing. It can be reduced by improving hydrodynamic conditions, i.e., reducing roughness or increasing oil viscosity or by using extreme pressure oils as in back axles where heavy loads at low speeds give ideal conditions for scuffing.

The normal type of "warm-scuffing" generates oil film breakdown by raising the temperature of the oil film locally to the point where the oil film can no longer maintain the surface apart. The exact mechanism is not known but research found that scuffing was associated with high friction power into the oil film (2,3). This Power input to the oil film depends on friction and on sliding speed and so is at a maximum where teeth take high loads well away from the pitch point. Once started the pattern of scuffing extends to cover the whole flank and gives a tearing of the surface running up or down the flank. Deterioration is progressive and little can be done after the initial stages of damage because the surface is so rough that it is difficult to re-establish an oil film. Cure of scuffing is usually by increasing the E.P. additive content though this can affect seal life.

Reduction in load is not usually possible and increase of diameter with the same module teeth gives a proportionate decrease in load only. Increase of facewidth helps and reduction of module helps so where possible both should be applied, leaving root stresses unaltered and contact stresses reduced. If it is possible to increase facewidth by 20% on a 6 mm module gear and reduce the module to 5 mm the product of load and sliding velocity has reduced by over a third.

Selection of oil viscosity is difficult as sometimes increase of viscosity giving a thicker oil film helps but decrease of viscosity can help by reducing heat generation as friction is lower.

IV. ROOT CRACKING

Tooth breakage is very uncommon. The trouble starts from small stress raisers in the root of the tooth propogating fatigue cracks which spread in a circular pattern from the "eye" defect and usually lead to loss of a large piece of the tooth or occasionally of the whole tooth. The result is often disastrous.

When inspecting gears, checking for root cracks is the most important part of the inspection since if any cracks are present, however small, the gear must be replaced immediately. Cracks may spread from pits on the flank itself to travel through the tooth but this is not common.

The causes of tooth breakage may lie in a tooth design that is overloaded, bad stress raisers in the root or in unsuspected overloads in the system due to starting or vibrations. Good inspection will help to stop root cracks in manufacturing but shot-peening of the root will greatly raise fatigue life. Tooth strength is controlled by module and so the possibilities are either to increase module, or increase facewidth or diameter to reduce loading. A check on T.E. will give an indication of whether dynamic loadings are giving trouble due to pitch errors.

V. WEAR

Wear involves steady removal of metal from the gear flank usually over the whole face and may either give the appearance of a lapped surface or may give a surface which is grooved in the sliding direction and so looks similar to a scuffed surface at first sight.

It can be caused by poor lubrication but this is unusual and it is debatable whether wear or scuffing occurs. The normal cause is the presence of abrasive in the oil; there is no inherent difference between the mechanism of lapping in gears to improve surface finish and wear of gears which produces mass metal removal. The cure is the removal of the abrasive which is causing the trouble; sand was found in one marine gearbox and had removed 60% of the volume of the teeth. Filtration of the oil should remove abrasive but there is rather a large difference between the 5 micron filtration which represents normal industrial practice and is suitable for hydrodynamic bearings and the micron film thickness that can occur in highly loaded gears.

In helical gears, thanks to the averaging effects, wear is fairly consistent across the whole face of the gear and so, although metal is removed, a roughly involute shape is maintained. The gears can lose surprisingly large amounts of metal without obvious effects on strength or the quiet running of the gears. The effects are very different with spur gears since the loadings are much higher during single tooth contact near the pitch point than when double contact is occurring near tip and root. Metal is removed roughly twice as fast in the center of the profile, rapidly giving a very high level of vibration; wear of only 100 microns is liable to require replacement.

In some industries the presence of abrasive is impossible to avoid on open gears. Printing and coal crushing are examples of this though modern design tends wherever possible to enclose gears to reduce noise level as well as contamination. Wear will occur and the objective of design is solely to reduce frequency of replacement; helical gears should be used whenever possible as they can give a life a factor of ten longer and the possibility of a local filtered air supply considered. It occasionally happens that very small pinions mesh with a large (and expensive) wheel and it may then be worthwhile recutting worn pinions, not to an involute profile, but to a profile which mates with the worn wheel thus economizing on wheel replacement.

VI. LESS COMMON PROBLEMS

Scoring is a variant on scuffing which occurs due to insufficient lubrication. The first part of the engagement, between driver dedendum and driven addendum operates satisfactorily but from the pitch point onwards the film is incomplete and scuffing occurs giving ridges in the sliding direction. The cure is simply to provide more oil, preferably by jet onto the flanks.

Plastic deformation is associated with heavy loads on soft materials. It may be beneficial to correct minor errors but if it occurs over much of the tooth face it is necessary to use a better material.

Case cracking occurs with drives such as worms and wheels where the worm is subjected to high sliding speeds which raise the temperature of the hardened case each revolution then quench it in cool oil. The resulting thermal fatigue cracks are usually radial. It may be necessary to reduce sliding velocities or to use a lower viscosity oil.

Dedendum attrition is a variant of pitting which occurs only below the pitch line. The cause is not known but it is thought to be associated with vibration and is cured by eliminating the vibration.

Electrostatic discharge occurs in machinery such as paper printing presses which can behave as electrostatic generators. The charges collected from the web travel to earth either via the bearings or via gears. Ball bearings have very low film thicknesses and so are a preferred path but if hydrodynamic bearings with relatively thick oil films are used the preferred path is through the gear contact at the pitch point where the oil film is at its thinnest. The resulting fine pitting is exactly on the pitch line. Cure is by an earthing brush.

REFERENCES

1. Neale M. J. Tribology Handbook. Butterworths 1973.
2. Blok H. The postulate about the constancy of scoring temperature. Interdisciplinary approach to the lubrication of concentrated contacts. (ed. P.M. Ku) NASA SP-237, 1970.
3. Winter H. and Michaelis K. Fresstragfähigkeit von Stirnradgetrieben antriebstechnik 14 (1975) pp. 405–409 and 461–465.

Chapter 14

LIGHTLY LOADED GEARS

I. EFFECTS OF LIGHT LOADING

Problems arise with gears when the applied load on the gears is not sufficient to maintain contact. The loss of contact is not harmful but the impacts involved when the teeth meet again can be damaging. This condition usually occurs due to light loads as when an engine is idling but can occur at high loads if the errors are larger than the tooth deflections and the system has high inertias.

The behavior of the system when it is non-linear is complex and we can get jumps and subharmonics as well as harmonics. Figure 14.1 shows a typical resonance curve for a non-linear system with a softening spring; as frequency rises there is a small jump to increased amplitude but as frequency decreases the resonance spreads over a large frequency range and eventually there is a sudden large decrease in amplitude. In the central region where there are two possible amplitudes of vibration; the response can jump from one to the other randomly as it is excited by pitch errors so the output is no longer regular and predictable.

As the system becomes more and more non-linear we can get the response shown in Figure 14.2 where the forcing T.E. (full line) produces a response (dotted) which is only in contact for less than 5% of the time. The impact occurs each time on a rising part of the T.E. and so each impact feeds energy into the vibration. In a more extreme case (with a lighter load) the impact occurs every second or third tooth so the response is a subharmonic of the forcing frequency. On a test rig this form of subharmonic vibration has occurred so strongly that every other tooth was badly damaged whilst the ones inbetween were untouched (literally).

Each individual impact as the tooth meet is highly impulsive and produces all possible frequencies so that if the impact force were analyzed over a short time, e.g., one tooth period it would give white noise at all frequencies, but when analyzed over a long period of time only multiples of once per revolution appear and there will be a strong content at twice and three times bounce frequency. In general the presence of strong harmonics of tooth frequency indicates loss of contact and if harmonics increase as load is reduced this is further reinforced; unfortunately this is not always true so it is not an absolute test.

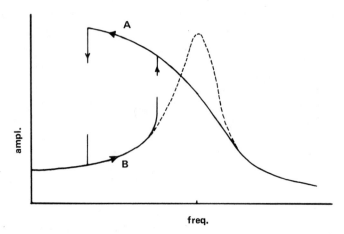

Figure 14.1.   Response curve showing jump effects. The dashed line is a linear system and the full line a non-linear system with a "softening" spring.

Extra complications arise when the system load is sufficiently light that the bounce takes up all the available backlash and impact occurs on the non-working flanks and the result is a powerful hammer across the backlash. This hammering may be regular or intermittent and often varies round the periphery of the gear. This often occurs in truck gearboxes at idling where excitation from the gear teeth is reinforced by torsional vibration from the engine to give loss of contact. Backlash hammer can damage gears but can also damage rolling bearings as the sudden velocity changes of a gear at impact can produce skidding in lightly loaded roller bearings with rapid fretting fatigue of the surfaces.

The main effects of loss of contact are not usually to damage the gears but to give very high vibration or noise levels and it is usually excessive noise at light loads that provokes investigations.

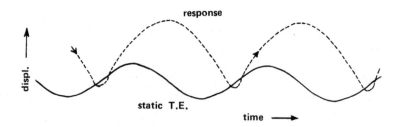

Figure 14.2.   Non-linear bouncing response. The full line represents the excitation, i.e., the T.E. and the dashed line the response as the gears bounce out of contact for most of the time.

II. DETECTION OF LOSS OF CONTACT

High noise levels at low loads, reducing as load is imposed, give a good indication that loss of contact is involved; it is not necessarily true as the curves of Figure 9.9, in Chapter 9, show. There is no absolute method of detecting whether loss of contact is occurring by vibration or noise measurement since there are exceptions to any "rule" that can be proposed. If speeds are low enough that T.E. can be measured, then it is possible to see whether loss of contact occurs only if vibration is predominantly torsional and not lateral; hence, even with T.E. there is not an absolute test for contact loss.

A very good indication occurs if "jumps" exist since these only occur with large non-linearities so any consistent difference between increasing and decreasing frequency responses is a strong clue. Alternatively a "warbling" or irregular changing between two levels of sound at given test conditions shows non-linearity. The presence of a subharmonic (half frequency) response is usually due to non-linearity at light load but may be due to time varying stiffness effects with spur gears at high loads.

Frequency analysis of noise or vibration is very little help since the results can look much the same for linear and non-linear systems. Signal averaging is slightly better since it shows up variations in contact conditions round the revolution; again similar effects can be obtained with badly pitched gears which are in contact.

As it is difficult, if not impossible, to say with certainty whether contact loss is occurring from tests especially where load cannot be varied, it is advisable to try some predictions.

III. CONTACT LOSS PREDICTION

Knowledge of whether separation is occurring or not is important because incorrect diagnosis of trouble will lead to "cures" which will make the vibration worse; in particular decrease of stiffness which will often help linear systems will correspondingly make a non-linear system worse.

The approach to prediction is generally that discussed in Chapter 10. Some idea of the system dynamics is required and a knowledge of the T.E. and the external excitation. Although the general case is inevitably complex some simple cases can be estimated rapidly and it is fortunate that these simple cases are ones which are likely to occur in practice.

At very high mesh frequencies, well above the natural frequencies of the system, the inertias cannot respond significantly to the once-per-tooth forcing and so at any instant the interference between the teeth is the mean tooth deflection under load less the transmission error. Taking a loading of 105 N/mm (600 lb.f/in.) facewidth on a helical gear with contact ratio 1.5 gives a loading on the teeth of 70 N/mm and a corresponding mean mesh deflection of 5 microns. A quasi-static T.E. of ±5 micron at once per tooth frequency would be needed to produce contact loss; this is unlikely on well made helical gears but very likely on spur gears.

At low speeds, system dynamics do not complicate the picture and the applied load must be large enough to accelerate the effective inertia of the system to maintain contact. Take a large wheel, effectively immobile, meshing with a pinion of radius 150 mm and moment of inertia 1 kgm^2 with a T.E. of ±5 micron at 180 Hz tooth frequency and assume solely torsional motion of the pinion. The angular acceleration is given by $5 \times 10^{-6}/0.15$ times $(2\pi \times 180)^2$ radians per second squared and the necessary load to maintain contact is then the acceleration times inertia divided by the radius which comes to 284N, a very low value. The acceleration involved at the teeth is 6.4 m/s^2 peak value and the effective inertia as seen at the teeth is $1/.15^2$ which is 44.4 kg; in most cases where only torsional vibration is involved it is convenient to work in terms of linear accelerations and effective masses (I/r^2) as seen at the teeth.

In complex cases a full analysis must be carried out to see whether the peak dynamic force exceeds the steady force level to give separation. Small amounts of separation are not important but when contact is for only part of the time as shown diagrammatically in Figure 14.2 it is worthwhile estimating the "flight" time each bounce to see if the loading will be on a up slope and, hence, the vibration will be energized or whether on a down slope which will extract energy.

A once-per-tooth T.E. of ±5 microns at a tooth frequency of 500 Hz will give a maximum "take-off" velocity of $5 \times 10^{-6} \times 2\pi x\ 500$ which is .016 m/s; additional eccentricity of 50 microns at 20 revs/s will give an additional .010 m/s, totaling .026 m/s. An effective gear mass at the teeth of 20 kg and a steady load of 600N combine to give an acceleration of 30 m/s^2 while out of contact. The flight time is then $026 \times 2/30$ which is 1.73 m/s and the maximum height is $0.5 \times .026^2/30$ which is 11.3 micron; the "flight time" assumes that the landing is at the same height as takeoff. In this case a flight time of 1.73 ms and a tooth frequency corresponding to 2 ms are sufficiently near that a small bounce on loading will increase the velocity enough to give a continued bouncing at once-per-tooth, probably for about two thirds of the revolution. The height of the bounce will then be up to about 20 microns and is not enough for there to be any danger of travelling across backlash of typically 150 microns.

As mentioned previously, damping is an unknown and so it is not possible to predict the energy loss in bouncing; the coefficient of restitution, i.e., the ratio of relative velocity after impact to the relative velocity before, will vary widely with size and natural frequencies as well as with lubrication conditions. For estimates of orders of magnitude we can assume arbitrarily that about two thirds of the energy is lost at impact so the coefficient of restitution, e, may be taken as 0.6. It is then possible for a gear mesh to exist stably in two different conditions according to whether it starts bouncing or not and it may jump from one condition to the other.

An example of this is given by two similar low helix angle gears, running in mesh with a contact load of 1,000N and with T.E. of ±5 micron at tooth frequency. The gears can only vibrate torsionally and each have an effective mass of 120 kg at the teeth. The total length of line of contact is 200mm. What is the maximum tooth contact frequency at which the gears can run before losing contact and at what frequency would you expect to observe maximum amplitude of vibration? It can be assumed that impact time is short, that shaft torsional stiffnesses can be neglected, and e is 0.6.

A displacement of ±2.5 micron per side and an acceleration of 1000/120 correspond to a frequency of 1826 rad/s or 291 Hz. The natural frequency of the two gears vibrating on a contact stiffness of $0.2 \times 1.4 \times 10^{10}$ is given by the root of $0.4 \times 1.4 \times 10^{10}/120$ which is 6,800 rad/s or 1087 Hz; as the natural frequency is well above the estimate of separation frequency, dynamic effects may be neglected.

Maximum amplitude will occur when the bounce time is exactly one tooth period. The T.E. gives a relative velocity of $5 \times 10^{-6} \times \omega$ so if bounce take off (and landing) relative velocity is V then $(V-5 \times 10^{-6}\omega) = 0.6 (V+5 \times 10^{-6}\omega)$. While out of contact, the acceleration of each gear is 8.3 m/s^2 and so the flight time is 0.5 × 2V/8.3 which equals tooth pitch time of $2\pi/\omega$. These expressions give ω as 1615, i.e., frequency 257 Hz. The closing velocity is .032 m/s and the bounce apart is 31.4 microns. This indicates that if the pair is run up above 291 Hz contact will be lost and the vibration will then increase steadily as speed is reduced until a maximum is reached at 257 Hz. On run up the vibration at 257 Hz will be 10 microns peak to peak but 31.4 microns on run down.

The assumption was made that the bounce was instantaneous; the actual time of bounce is very nearly half a cycle of the natural frequency whilst in contact since the steady load is low and, hence, static deflection is low. The natural frequency of 1087 Hz gives a contact time of 0.46 milliseconds compared with the estimated 3.89 milliseconds tooth period so the estimate will be about 10% out and maximum height will be achieved at slightly lower speed. Of interest also is the maximum loading on the gear teeth; the closing velocity is .032 m/s so each tooth must deflect an amount corresponding to .016 m/s at a frequency of 1087 Hz and each deflection is .016/6,800 which is 2.4 microns. The total mesh deflection of 4.8 microns then gives a peak load of $4.8 \times 10^{-6} \times .2 \times 1.4 \times 10^{10}$ which is 13,440N, some 13 times the nominal load.

IV. PREVENTION OF BOUNCE

Curing loss of contact is rarely easy. The obvious solution of reducing T.E. is sometimes possible with helical gears but rarely with spur gears. The only other variables which can be altered are inertia, stiffness, damping, and applied load.

Reduction of inertia or increase of applied load both increase the frequency at which bouncing will start and so will help for systems below resonances. Above resonances, changing inertia has little effect though increase of load increases steady tooth deflection. The overall effect on stressing is small but the harmonic content of vibration will be reduced.

Reduction of shaft stiffness is liable to make separation worse at low frequencies and so increase impact noise but is unlikely to have any effect well above resonance. Reduction of tooth mesh stiffness, provided it does not move resonances into working ranges, always seems to help at both high and low frequencies. It is probably true to say that reduction of mesh stiffness helps both linear and non-linear systems for noise though there is inevitably a root stressing penalty associated with more flexible teeth. In cases where loads are very low plastics greatly reduce mesh stiffness.

Damping is always helpful but is difficult to apply since a torsional or lateral vibration absorber must be fitted to the body of the gear according to the mode of vibration. Tuned absorbers are of little use since frequencies cover wide ranges with non-linearities and so the Lanchester type of viscous or friction coupled damper must be used. Fitting a damper will not usually prevent loss of contact but will greatly reduce the amplitude of vibration and hence the power in the impacts. Choice of damper settings can best be done by setting up a digital or analog model of the system.

When hammering across the backlash occurs there is an additional variable in the amount of backlash specified. The level of vibration and noise on a system reaches a maximum with the amount of backlash that corresponds to synchronism of bounce and forcing T.E. Testing at a fixed speed does not allow investigation of whether speeding the bounce up or slowing it down gives an improvement and so the only solution is to try different levels of backlash or to model the vibration. Unfortunately changing a gear pair usually changes T.E.; the ideal solution is to test a gear pair at variable center distance. Prediction of separation distance as in Section III will help since this will show whether it is possible to increase backlash to the point where no hammer occurs.

Chapter 15

GEAR SYSTEMS

I. USES OF GEARS

Gear usage can be classified under four main headings for parallel shaft gears as use for synchronization, speed step-up or down, variable ratios, or for power addition.

Synchronous drives, whether by gears, toothed belts, or chains are necessary when exact position relationship must be maintained as with overhead cam drives in engines, color register in printing presses, and needle drives in knitting machines. Powers are often low in relation to the size of the gears and the main requirement is high accuracy, i.e., low T.E. Noise problems with this type of drive are often associated with low torque or reversal of torque giving movement across the backlash and then the amount of backlash must be kept low.

Variable ratios are required for machine tools and car gearboxes to give matching between a power unit whose speed is fixed or cannot vary greatly and a load which may vary speed over a 100:1 range. In a wide ratio gearbox it is customary to give speeds on a geometric progression with a ratio of about 1.4 so that speed doubles every other gear as this allows selection within 20% of speed over the whole range. The method of cascading gear ratios is sketched in Figure 15.1 and the overall ratio is the product of the ratios for each stage. Choice of ratios for the stages is more complex than it appears since no two stages should give the same overall ratio which is wasteful. A three stage box with ratios of 1 or 1.41, 1.41 or 0.71, and 2.83 or 0.71 gives 8 ratios from 0.5 to 5.63 in geometric progression. Design is soon restricted by the limit of about 5 to 1 reduction that can be achieved easily in a single stage. This type of gearbox is prone to lateral vibration of the gears since shafts are relatively long and flexible in bending as well as torsionally.

Power addition in its simplest form is used when more than one engine or crankshaft drive a common load; a typical example is the use of two turbines driving a single propellor. Two (or more) pinions mesh with the same wheel to give addition of torque. In this type of power addition, relative speeds remain unaffected and torques add. If the requirement is for torques to remain constant but speeds to add then a differential or epicyclic system must be used.

147

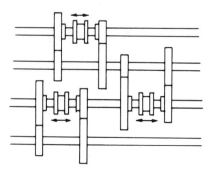

Figure 15.1. Sketch of eight speed gearbox using three two way clutches with three pairs of meshes.

Speed reduction using parallel shaft gears is straightforward; problems arise in the economic balance between increasing number of stages and increasing ratio per stage. In practice 5 to 1 is the usual limit for a single stage though 10 to 1 is possible and an overall ratio of 60 to 1 would require a detailed costing to choose whether 2 or 3 stages would be better. High ratios of the order of 60 to 1 are easily achieved using worms and wheels in one stage but this type of gearbox is not suitable for high powers or high speeds. Step-up gear drives present similar problems but worms and wheels are not used.

II. TORQUE SPLITTING

In large installations running at low speeds, such as marine drives, the output torque is very high and a single output mesh drive becomes large. 15,000 kW (20,000 HP) and a propellor speed of 120 r.p.m. would require a torque of 1.2 $\times$ 10^6 Nm; assuming a facewidth of 0.75m and a loading of 500 N/mm (3000 lb./inch) facewidth gives a wheel base diameter of 6.4m, i.e., a pitch diameter of 6.8m. Splitting the power into two pinions driving onto a common wheel with the same facewidth and loading would halve the diameter giving a sensible size. Splitting onto 4 pinions would further reduce the size of the final wheel (but would also reduce the final drive ratio).

It is uneconomic to use 4 separate turbines and so the tendency is to split the power from a given turbine into two intermediate speed shafts so that each turbine drives 2 final pinions, as in Figure 15.2. There will inevitably be errors in the gears and so a rigid drive cannot be used; some form of flexibility must be designed into the drive to ensure that equal forces go through the two meshes despite errors. Maintaining load sharing within 5% on a double drive with possible manufacture and build errors of at least 100 microns requires a static load deflection of the flexible equalizing system equivalent to 2mm at the pinion teeth; this is easily obtained with a torsionally flexible drive shaft, though other

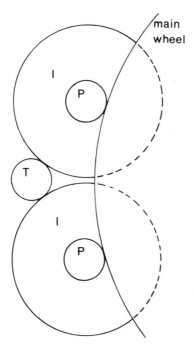

Figure 15.2.   Method of splitting power onto two final drive pinions P from a single turbine shaft T. The two intermediate line wheels I are coupled to the pinions P by torsionally flexible shafts.

methods are possible. The existence of low torsional stiffness can help to prevent vibration transmitting between the first and second stage and so usually reduces transmitted noise; the effect on natural frequencies must be checked.

An alternative form of load sharing for high torques is the epicyclic gear which is extremely compact.

III. EPICYCLIC GEARS

An epicyclic gear is shown sketched in Figure 15.3. It is possible to keep the planet carrier stationary (usually termed a star gear) but normally the annulus is fixed as this gives a greater reduction ratio and a higher output torque for a given size of gear. There are many different reasons for using epicyclic gears despite the greater complexity, lower accessibility, and high bearing loads involved.

Epicyclic gearboxes are used in automatic car transmissions because gear changes can be achieved with external powerful brakes which have sufficient space to allow room for generous friction surfaces. For some installations the coaxial input and output shaft is an advantage and occasionally the fact that the

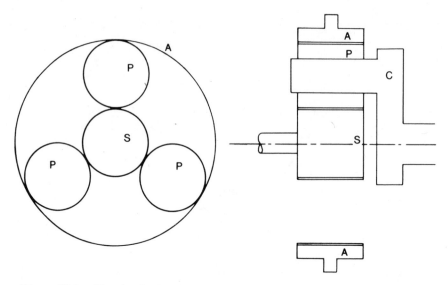

Figure 15.3.   Sketch of epicyclic gear. S is the sun wheel, A the annulus, C the planet carrier, and the planets are labelled P. Support bearings are not shown.

gearcase reaction is a pure torque about the axis can be used to measure system torques or to provide very effective vibration isolation. A dominant advantage of epicyclics is the high torque-to-weight ratio and the very compact layout that they can achieve.

Load sharing in parallel shaft gears gave typically four tooth loads at wheel radius. This is matched by an epicyclic as in Figure 15.3, because there are three tooth loads at annulus radius and further three tooth loads at sun radius which is about a third of annulus radius; this is equivalent to four tooth loads at full radius. The epicyclic gear is, however, more compact than the comparable parallel shaft gear because extra space is not used on the pinions. If a single turbine is used, only two pinions are usually used in a parallel shaft installation and sizes are correspondingly large whereas with an epicyclic it is possible to go to a 5 planet design which gives the equivalent of seven tooth loads at annulus radius though with a reduction ratio restricted to about 4 to 1.

Epicyclic gears are typically about a half or a third of the volume and weight of comparable parallel shaft gears but are more difficult to manufacture and have high planet bearing loads in restricted room. They are in general less accurate than their rivals but can be vibration-isolated more easily.

Calculation of gear ratios can be carried out simplest by the standard tabular method, initially fixing the planet carrier and determining system speeds, then superposing a rotation on the whole system to bring any other member to rest or to a specified speed. A good example is an epicyclic gearbox as used for

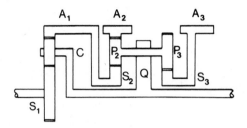

Figure 15.4. Coupled epicyclics for automatic gearbox. S_1, S_2, and S_3 are the sunwheels; C the first stage planet carrier is coupled directly to Q, the carrier for the second and third stages and the output shaft. A_1, A_2, or A_3 may be held by band brakes to give the different ratios.

automatic transmission, see Figure 15.4. Tooth numbers are A_1 and A_2 80, S_1 and S_2 40, and S_3 30; what is the ratio for A_1, A_2, or S_3 held fixed? All teeth are the same module.

Tooth numbers, $S_2 + 2P_2 = A_2$ so P_2 has 20 teeth and as S_3 has 30 teeth P_3 must also have 30 teeth. Then:

	A_1	C	S_1	A_2	Q	S_2	P_2	S_3	Q	P_3
	$-\dfrac{n}{2}$,	0,	n,							
				$-\dfrac{m}{2}$,	0	m	$-2m$			
								r	0	$-r$
Fix A_1	0,	n,	$\dfrac{3n}{2}$,	$-\dfrac{3m}{2}$,	$-m$	0				
Fix A_2	x,	$\dfrac{x+n}{2}$,	$\dfrac{x+3n}{2}$,	0,		$\dfrac{m}{2}$	$\dfrac{3m}{2}$			
Fix S_3	$\dfrac{y-n}{2}$,	y,	$y+n$,	$\dfrac{y-m}{2}$,	y	$y+m$,	$y-2m$,	0	$-r$	$-2r$

A_1 fixed gives ratio of $\dfrac{3n}{2} \div \dfrac{n}{2}$, i.e., 3:1 reduction.

A_2 fixed $x = 3m/2$ and $x + n/2 = m/2$ so $n = -2m$

and so ratio is $x + 3n/2 \div m/2$ which is $-3:1$ reduction (reverse)

S_3 fixed $y - 2m = -2r$, $y + m = y - n/2$, $-r = y$

so that $n = -2m$, $r = 2m$ and ratio is $(y+n)/y$ which is $-4m/-2m$, i.e., 2:1 reduction. Clutching S_1 to A_1 (as used in the Wilson box) locks the epicyclics to give 1:1 through drive.

It was mentioned earlier that an epicyclic gear can be used as a dynamometer to measure torque in a system. The external torques on the gearbox must sum to zero and the other essential relationship is derived from the efficiency of the system. In the gearbox above, with A_2 fixed, what is the reaction torque when the input torque is 90 Nm and the efficiency is 95%?

Velocity ratio is –3 so output torque is $0.95 \times 3 \times 90$ which is 256.5. If the input torque is 90Nm clockwise the output torque is 256.5 Nm anticlockwise so the externally applied torque is 256.5 clockwise and, hence, the restraint torque is 346.5 Nm anticlockwise.

The design of epicyclic gears with standard teeth requires that the number of teeth in the annulus less the number of teeth in the sun must be even and if the planets are equally spaced the number of teeth in sun and annulus must be divisible by the number of planets. Both rules can be "bent" by using non-standard teeth and by slight movement of the planet centers.

An unusual type of epicyclic called an harmonic drive uses a rigid annulus meshing with a flexible annular inner which has two less teeth. Rollers distort the inner to an elliptical shape so that it contacts at two points only as shown in Figure 15.5. A complete revolution of the beam carrying the rollers gives a rotation of the inner of only two teeth so reduction ratios of the order of 100:1 are achievable. A variant on this is to maintain the distorted shape by magnetic coils which switch to progress the distortion round the annulus.

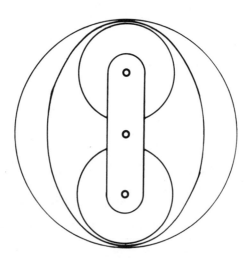

Figure 15.5. Principle of harmonic drive. The flexible inner annulus has two teeth less than the rigid outer annulus.

IV. EPICYCLIC LOAD SHARING

At high torque levels the great advantage of the epicyclic lies in its load sharing
so it is very important that the load is shared evenly. There are a very large
number of load sharing systems, all of which are claimed to be the "best;" it is
advisable to assess them from both static and dynamic aspects.

Taking a reasonable accuracy of manufacture and build of, say, 50 microns
for static and one-per-revolution effects and a requirement for balance within
10% gives a steady state deflection of 500 microns (.020 in.) necessary to accommodate errors. Some manufacturers have claimed that their gearboxes were so
accurate that load sharing devices were not required but this claim is unlikely.
The methods used fall into categories of:

a. floating the annulus,
b. floating the sunwheel,
c. linkages,
d. flexible annulus and
e. flexible planet pins.

Allowing the annulus to move radially is effective in equalizing loads on
three planet designs as equilibrium of the annulus (neglecting its weight) dictates
that the three forces must be equal. The annulus is a weighty component and so,
although this system would work well statically, it is poor dynamically. The sun
wheel is much lighter than the annulus and can be allowed to move radially very
easily; it gives good load sharing but is restricted to three planet gears.

Dynamically it is very much better than a floating annulus but cannot respond at high frequencies.

Designs which use linkages between the planets to give load sharing will
work adequately statically but the inertias of planets, rigid supports, linkages,
and equalizing mechanisms are sufficiently large that it is difficult for the system
to respond to high frequency excitation.

Flexure of the annulus itself or of the planets pins gives the most successful approach to load balancing which is also effective at high frequencies. In
both cases elastic deflections are large, typically 750 microns tangentially on the
planet pin or a similar amount radially on an annulus ring; due to the meshing
geometry the annulus ring must deflect radially rather more than the required
meshing error.

Comparison of the annulus versus planet pin approach is not straightforward. Flexing the annulus involves moving less mass than moving a complete
planet but the comparable movement involved may be three times larger and so
effective mass is a factor of 9 up. Errors between planet and annulus are easily
accommodated with annulus flexure but sun-to-planet errors would have to

transmit dynamically through the planet inertia to reach the annulus. This will not happen because it is easier for the errors to move the sun wheel and so the annulus will have little effect on high frequency errors at the inner mesh. Planet pin flexure will allow errors at either inner or outer meshes to move the planet .tangentially. Annulus movement radially may be limited in practice by the geometric effects on the mesh as the profile is tilted by the annulus distortion but this tilt can be allowed for in the manufacture. In both cases detailed checks on natural frequencies are required to ensure that natural frequencies of the system do not lie near tooth frequencies.

V. SPLIT POWER SYSTEMS

As energy prices rise there is increased demand for high efficiency in drives and for systems which are capable of storing energy. Both requirements can lead to a need for power addition involving gear drives.

High efficiency can be maintained in a vehicle engine if it is run at constant speed but it is then necessary to have a infinitely variable drive between engine and wheels. Whether mechanical, hydraulic, or electrical variable drives are used the efficiency is poor with losses ranging from 10 to 20% with present designs. One method of reducing the effects of inefficiency is to use the type of circuit shown in Figure 15.6. The infinitely variable drive, usually hydraulic, is capable of absorbing or returning a fraction α of the input engine power. The engine is connected directly to the sun of an epicyclic and the annulus is driven by the variable drive to increase or decrease output speed. Neglecting losses, the speed ratio at output will vary from $1 + \alpha$ down to $1 - \alpha$ times locked-annulus speed. When the annulus is locked, losses in the system should be low because only idling losses are involved and mechanical efficiency is high. At top or bottom output speed the losses will only be a fraction α of the losses that would

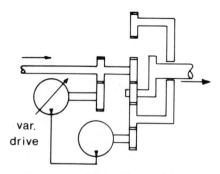

var.
drive

Figure 15.6. Split power system. Part of the input power may be fed via the variable hydraulic section to speed up the output by driving the annulus or power may be returned from the annulus to input, reducing output speed.

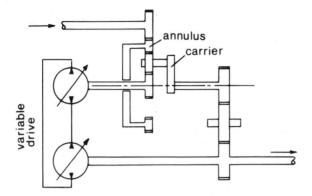

Figure 15.7. Split power system as used by Vaughan at Bath. Pump and motor
are both variable and speed can reduce to zero.

be involved if the variable drive took the full power. There are several possible
variations on this type of drive and it is often necessary to have separate arrangements for very low speeds and reverse with extra clutches. Figure 15.7 gives
another variant which is suited to vehicle work but which requires both hydraulic pump and motor to be variable delivery.

Infinitely variable systems are not yet competitive and a simple manual
gearbox is much cheaper, efficient and more economical because it has low
mass and this aids economy. The tendency towards 5 speed gearboxes will increase as engine sizes decrease and at the moment the most economical combination in a vehicle is a medium size (1 to 1.5 liter engine) with a five speed
gearbox to give very low engine speeds at cruising.

This combination gives 44 miles per U.S. gallon (55 miles per British
gallon) for a car which can carry four adults cruising at 56 m.p.h.; top speed is
then typically 75 m.p.h.

The most likely use of split-power systems in the future will be in large
earth moving machinery where the controllability of hydraulic systems is essential for part of the work cycle, though powers may then be low but the inefficiency of hydraulics is a large penalty. The combination of hydraulic and
mechanical drive with possibly two separate engines can give large fuel savings.

VI. ENERGY STORAGE

The advantage of having an energy storage system is that a small engine with low
losses can be fitted and that regenerative braking can reclaim energy used for
acceleration. If only short trips are made in cars the energy storage system may
be sufficient by itself.

A typical family car requires about 20 kW for steady cruising but is fitted
with a 75 kW engine for occasional bursts of acceleration or hill climbing. A

single acceleration of 1000 kg of car to 20 m/s (45 m.p.h.) requires an energy of 55.5 Watt-hours or 200 kWs (i.e., kJ). Lead acid batteries normally hold about 40 Watt-hours/kg and so a single small battery holds enough energy for one acceleration or to climb a 20m high "hill." The problem is that although batteries contain a lot of energy they cannot give out or re-absorb energy faster than about 0.5 kW per battery without damage. A large number of batteries must be used to give reasonable power levels and the weight penalty is then large; however, once this weight of battery has been installed to give the necessary power rating the range of the vehicle is greater than 50 miles on the battery capacity above.

The pattern of the future, if liquid fuels become expensive, is likely to be a small engine, perhaps 10 kW only, used solely for long trips with a battery storage which is sufficient for commuting and will also work in parallel with the engine for acceleration. Electrical storage is the most likely method since although electrical storage at 120 kJ/kg is poor compared with liquid fuels at 40,000 kJ/kg it is better than mechanical (flywheel) energy at 16 kJ/kg, gas pressure energy at 8 kJ/kg or spring energy at 0.03 kJ/kg.

With all these systems there is likely to be a requirement for gearing to add in power to a back-up engine in a parallel drive system, so some form of epicyclic or differential is likely.

APPENDIX

UNIT EQUIVALENTS

The S.I. system is based on the meter (m), the mass of 1 kilogram (kg) and the second and involves forces in Newtons, 1 N being defined as the force required to accelerate 1 kg at 1 m/s^2.

The unit of work is the Joule (J) which is the work done when a force of 1 N does work a distance of 1 m. The dominating advantage of the S.I. system is that the unit of work 1 Joule, which is 1 Nm, is also the unit of electrical work 1 ampere through 1 volt. Power is in J/s which is termed Watts. The basic equivalents are:

1 inch	= 25.40000 mm	1 m	= 39.3701 inches
1 foot	= 0.3048 m	1 micron (μm)	= 0.04 thou (mil)
1 lb	= 0.453592 kg		
1 lbf	= 4.44822 N	1 ft. lb f	= 1.3558 J (work)
1 hp	= 745.700 W	1 lb f ft.	= 1.3558 Nm (torque)

Derived units with common typical figures relevant for gearing work are then:

1 lb.f/in.^{2} = 6894.8 N/m^2 30,000 psi   = 2.07 × 10^8 N/m^2
(Pressure or modulus)   or 207 N/mm^2
                                 30× 10^6 psi   = 210 × 10^9 N/m^2
1 lb.f/in. = 175.13 N/m   10^6 lb.f/in.   = 1.75 × 10^8 N/m
1 lb.f/in./in.= 6894.8 N/m/m 2× 10^6 lb.f/in./in. = 1.4 × 10^{10} N/m/m
(Stiffness per unit facewidth)   or 3.5 × 10^8 N/m/25 mm
1 P (poise) = 0.1 Ns/m^2
1 St (stokes)= 10^{-4} m^2/s   1 centistoke = 10^{-6} m^2/s
1 in^2/s = 645.16 centistokes (kinematic viscosity)
1 lbf.s/ft^2 = 47.88 Ns/m^2 (dynamic viscosity)
1 lb.ft.^{2} = 0.04214 kgm^2
1 slug ft.^{2} = 1.3558 kgm^2 (moment of inertia)
1 lb.in.^{2} = 2.9264 × 10^{-4} kgm^2

Though not part of the S.I. system, the kilopond is used in Europe as a measure of force and is the weight of 1 kilogram and so is 9.80665 N. Pressures are often expressed as bars, 1 bar being 10^5 N/m^2 and approximately 1 atmosphere; 1 atm exactly is 1.01325×10^5 N/m^2. There is no general agreement on whether values of, say, modulus should be expressed as 210×10^9 N/m^2 or 210 kN/mm^2 but is is probably safer to use the first method in all calculations.

Typical values for steel for modulus are $E = 210 \times 10^9$ N/m^2, $G = 81 \times 10^9$, $K = 175 \times 10^9$, Poissons ratio 0.3, density 7843 kg/m^3.

PROBLEMS

Chapter 2

1. A wheel and pinion have 60 and 25 spur teeth, and both gears have an addendum equal to the module of 2.5 mm. What is the center distance and what are the tip diameters?

What is the contact ratio if the pressure angle is
a) $14\frac{1}{2}°$
b) $20°$
c) $25°$?

2. Two spur gears have stub teeth for high strength. They each have 32 teeth with a pressure angle of $20°$. What must the addenda in terms of the module be if the contact ratio is not to be less than 1.10?

An alternative design uses 32 tooth gears with standard addendum of the module but must have a contact ratio of 2. What is then the pressure angle?

3. A wheel and pinion have 50 and 30 spur teeth with $20°$ pressure angle and 3 mm module. The addendum equals the module. What is the length of action and what is the contact ratio? What is the relative radius of curvature for contact at the pitch point?

4. In a rack railway the pinion has 25 teeth of 20 mm module with $20°$ pressure angle and rotates at 100 rev/min. If the addendum for rack and pinion is equal to the module, what is the contact ratio and what is the sliding velocity as the teeth come into contact? What is the instantaneous friction power if the load is 20,000 N and the coefficient of friction is 0.1?

Is it possible to increase the contact ratio to 2 so that there are always two pairs of teeth working by increasing the rack addendum only?

5. A wheel of 24 teeth, module m, and $20°$ pressure angle meshes with an annulus of 72 teeth. If the wheel addendum is m and the contact ratio is 1.50, what is the annulus addendum? What is the minimum value of the relative radius of curvature?

6. Two gears each have 24 teeth of standard shape with $20°$ pressure angle. What is the contact ratio when they are meshed at the correct center distance?

The gear axes are then moved apart until the contact ratio is 1.00. How far must the axes be separated (in terms of m) and what is then the operating pressure angle?

7. a) For a pair of gears of module 6 mm and with axial breadth 40 mm, the safe load is 200 kW when the driving shaft speed is 1000 rev/min. Estimate the safe load for a pair of gears with a geometrically similar transverse section and of module 4 mm, when the driving shaft speed is 2000 rev/min, the axial breadth is 25 mm, and the same contact stresses can be allowed. Neglect dynamic effects.

b) A gearbox diameter and length are increased by 30% but the gear tooth size, i.e., the module remains unchanged. What is the permissible increase in torque if contact stresses limit and if root stresses limit?

8. A spur gearbox will transmit 1200 kW at 180 rpm using 5 mm module gears. A new design is required to transmit 1600 kW at 150 rpm. Assuming that a similar design of gearbox is used and that dynamic effects can be neglected, what are the possible designs, using 6 mm module gears with the same facewidth so that axial length is not increased.
a) contact stresses limit?
b) if root strengths limit?

Chapter 3

1. A pair of helical gears are to have 3 mm m_n teeth of standard form, $20°$ pressure angle with 24 and 75 teeth and a pitch helix angle of $25°$.

What will be the center distance and the transverse contact ratio? If the facewidth is 35 mm and the permissible load on this size of tooth is 125 N/mm based on the average length of line of contact, what is the torque on the 75 tooth gear?

What is the minimum length of the line of contact and what helix angle would you suggest to improve the design?

2. A pair of gears is to be cut with a hob designed for module 4 mm and pressure angle $20°$. The wheels are to have 24 and 40 teeth and to work between parallel shafts 140 mm apart. Find the appropriate pitch helix angle, transverse pressure angle, and base helix angle.

If for each wheel the axial thickness is 30 mm and the addendum is equal to the normal module, find the mean force per millimeter of contact line when the gears are transmitting 80 kW at speeds of 3000 and 1800 rev/min. Neglect losses.

3. A pair of gears with 65 and 25 teeth, 20° normal pressure angle and standard tooth dimensions are to work between axes 160 mm apart. If the normal module is a whole number of millimeters find m_n, the pitch helix angle, the transverse pressure angle, and the contact ratio.

The axial width of the gears is 40 mm. What are the minimum, mean, and maximum length of the line of contact and what is the pinion torque based on mean length of line if the permissible load is 200 N/mm?

4. What is meant by the virtual spur wheel of a helical gear wheel? A pair of steel helical gear wheels are required to be cut with standard cutters, to have 40 teeth each, and to transmit power between shafts 100 mm apart. Find the largest acceptable cutter module and the corresponding pitch helix angle. Determine the radius of the virtual spur wheels and the relative radius of curvature of tooth elements in contact a) at the pitch point and b) just as contact begins or ends. For each condition calculate the permissible load per unit length of contact line if the maximum compressive stress in the steel is limited to 500 N mm^{-2}.

(Standard modules in this range are in 0.25 mm steps, i.e., 2, 2.25, 2.5, etc.)

5. A wide pinion has a fixed pitch circle diameter and meshes with a very large wheel. It has 30 teeth of normal module 8 mm, ϕ_n of 20°, and a helix angle of 4°. To improve smoothness of drive the helix angle is to be increased to about 30° without altering pitch radius or m_n. What will be the new helix angle and the number of teeth on the pinion?

Determine the radius of curvature at the pitch point for the original and modified designs (using the Gregory method) and, hence, deduce the change in permissible torque if surface contact stresses are unaltered.

6. Two mating helical gears have 23 teeth of 4 mm module with a pitch helix angle of 40° and a facewidth of 55 mm. Assuming that load is controlled by root stresses, is proportional to minimum length of line of contact and to module and that at least 23 teeth are required, how would you increase load capacity without altering materials, center distance, or facewidth? What percentage increase of torque would you achieve?

7. Two possible designs of pinion are being considered for a rack drive, using 22 teeth 8° helix angle and 10 mm module standard gears or using 24 teeth 35° helix angle 10 mm module standard gears. The facewidths are sufficiently large that average contact line lengths may be used. What is the ratio of permissible drive *forces*
   a) If surface stresses limit?
   b) If root strength limits?
Neglect effects of tooth profile changes on root stresses.

8. A "balanced" gear design is one in which root and surface stresses reach their limiting value simultaneously. A design is limited to a surface contact stress of 500 N/mm^2 and to a root bending stress of 300 N/mm^2; for the $20°$ pressure angle tooth shape chosen the geometric factor (Y_{FS}) (stress at $30°$ point times module divided by load per unit facewidth) may be taken as 4.75.

What number of teeth on a spur gear will give a balanced design when the gear mates with:
   a) A similar gear?
   b) A rack?

Chapter 7

1. A spur wheel and pinion transmit 200 kW at a pinion speed of 2,900 rpm. The pinion facewidth is 50 mm and its base circle radius is 70 mm. Alignment on assembly is such that overload due to helix effects is less than 100%, the *adjacent* pitch error on each gear is kept below ±3 microns and the profile is consistent to within ±1 micron.

What tip relief would you suggest? Assume no root relief and equal tip relief on wheel and pinion. If the wheel is large and the pinion has 24 teeth, how far in terms of roll angle from the pitch point should the tip relief finish?

2. A pair of spur gears have 75 mm facewidth and carry a load of 12,000 N. The helices are each held to an accuracy of ±5 seconds of arc and the gearcase housing alignments are held to ±20 microns in 300 mm *between* the shafts. Casing and shaft deflections under load may be neglected. What is the maximum loading per unit facewidth that can be encountered neglecting dynamic effects?

3. A gear has base circle diameter 152 mm, pitch diameter 168 mm, root diameter 156 mm, and 200 mm facewidth. The imposed load is 80,000 N. It is mounted symmetrically on bearings whose centers are 400 mm apart and is driven from one end only.

What corrections to the helix would you suggest?

4. A gear carries a total load of 30,000 N and is itself rigid but is carried on a shaft of 100 mm diameter overhung 90 mm to its center. The shaft bearings are spaced 500 mm apart and the shaft is 130 mm diameter between the bearings which have a stiffness for lateral loads of 2×10^9 N/m (10^7 lb.f/in.). What correction should be applied to the gear helix?

5. A spur gear is supported between bearings on a shaft 80 mm diameter and its center is 300 mm from one support bearing and 500 mm from the other bearing. A tooth load of 5600 N is applied and bearing stiffness is 10^8 N/m.

What angular correction should be applied to the gear to compensate for shaft bending? Assume the mating gear is rigidly mounted and has no corrections.

6. A pinion has an effective diameter of 200 mm and 250 mm facewidth and meshes with a large wheel. It is mounted symmetrically between bearings 590 mm apart and has a loading of 600 N/mm on its facewidth. Wind-up effects may be ignored.

Estimate the pinion distortion due to bending. If the operating transverse pressure angle is 22° and the wheel rim is only supported at its center, how much should the rim distort at its edges under a radial line load of 100 N/mm applied across the facewidth.

Chapter 8

1. An idealized model of part of a gearbox (as in Figure 8.3) consists of a pinion of mass 5 kg and radius of gyration 40 mm, base circle radius 50 mm, supported on a shaft with lateral stiffness 2×10^7 N/m, and a torsional stiffness of 3×10^5 Nm/radian meshing with a wheel of mass 40 kg and 120 mm radius of gyration, 140 mm base radius supported on a shaft stiffness of 10^9 N/m with torsional stiffness 5×10^6 Nm/radian. The facewidth is 60 mm and the contact ratio is 1.5. The effects of helix angle on mesh stiffness may be neglected and it may be assumed that large inertias prevent twist at the ends of the shafts.

Estimate the tooth contact forces and the forces transmitted to the bearings per micron of meshing error at frequencies of 10 Hz, 250 Hz, and 2400 Hz. Neglect damping.

2. A simple single mesh system (as in Figure 8.3) has parameters $r_b = 45$, M = 3 kg, J = 0.03 kgm², S = 2×10^8 N/m, K = 4×10^5 Nm/rad for the pinion and the wheel is massive so its vibration may be ignored. The mesh stiffness is 10^9 N/m.

What are the forces per micron error transmitted to the bearings at 50 Hz and at 1500 Hz? How would the transmitted forces at 50 Hz and 1500 Hz be altered by a design change which reduced the contact stiffness to 5×10^8 N/m?

3. Compare the expected vibration levels from two possible designs for a low power high speed drive where the design figures are
   a) Pinion 1 kg, J = .005 kgm², S = 10^7 N/m, K = 10^4 Nm/rad, r = 25 and wheel 5 kg, .08 kgm², 2×10^8 N/m, 3×10^5 Nm/rad, r = 60 with steel gears 15 mm facewidth and 1.5 contact ratio and expected errors of ±3 microns at 2000 Hz.
   b) A drive with nylon wheel and steel pinion, with the same inertias and stiffnesses except for the contact stiffness which is reduced by a factor of

15 (as facewidth is increased) and with expected errors of ±8 microns at 2000 Hz.

Chapter 10

1. For the system of problem 8.1, determine the natural frequencies by setting up the equations in matrix form.

2. For the two systems of problem 8.2, determine the natural frequencies by a matrix approach.

3. For the systems of problem 8.3, determine the natural frequencies by a matrix approach.

4. At a gearbox foot the local response to vibration is measured as 2×10^{-8} m/N lagging at $45°$ phase, at 500 Hz; when the gearbox is running the amplitude of vibration is 0.2 micron.

   A mass of 10 kg is rigidly bolted to the foot. What vibration level would you then expect when the gearbox is run.

5. A gearbox has a single wheel-to-pinion mesh and is tested by applying a vibrator to the wheel and to the pinion separately with the teeth out of mesh. The observed test results are tabulated, with 100 N test force in each case at tooth frequency in the direction of the line of thrust.

Force	Response wheel	Response pinion	Vibration at foot
On pinion	0.1 micron lagging 135°	1 micron lagging 35°	0.1 micron at 250°
On wheel	0.2 micron lagging 70°	0.1 micron lagging 135°	0.05 micron at 160°

What transmitted vibration at the foot would you expect for a 1 micron error at tooth frequency and what would the tooth force be?

6. A perspex model of a gearbox is made, one quarter of full size. Perspex has a modulus of 3×10^9 N/m^2 and a density of 1200 kg/m^3. Tests on the model give a natural frequency of 620 Hz and a local stiffness of 5×10^6 N/m.

   What are the expected natural frequency and local stiffness of the full size gearbox in steel?

Chapter 11

1. A gear mesh has a wheel of mass 30 kg with 90 teeth, base radius 206, polar inertia of 1 kg m^2 and restraint stiffnesses of 4×10^8 N/m laterally and 10^6

Nm/rad torsionally. The pinion mass is 5 kg with 24 teeth, polar inertia 0.08 kg m^2, base radius 55 mm and stiffnesses 1.4×10^8 N/m laterally and 1.5×10^5 Nm/rad torsionally; contact stiffness is 10^9 N/m.

When the wheel speed is 400 rpm what are the laterial forces transmitted to the bearings per micron of T.E. at tooth frequency? If the mean tooth load is 12000 N suggest possible changes in shaft and bearing support stiffnesses to reduce the transmitted force. It may be assumed that a relative centerline deflection of 0.3 mm can be allowed.

2.   It is proposed to use coil springs to isolate a gearbox and motor. The springs when in position have a length of 0.1 m a mass of 0.2 kg and a stiffness of 40,000 N/m and each spring supports a mass of 40 kg. What are the natural frequencies of the system for vertical excitation?

3.   A gearbox has a mass of 400 kg and can be regarded as rigid. It is mounted using isolators on a subplate which with auxiliaries has a total mass of 600 kg and the subplate is isolated from the (rigid) floor. Only vertical vibration is of importance and the relevant excitation frequencies are 5 Hz, 24 Hz, 300 Hz, and 2000 Hz. Applied vertical force to the gearbox can be about 10,000 N and the gearbox must not move more than 0.5 mm when this force is applied.

What isolation would you suggest and what ratio would you expect between vibrating force and force transmitted to the floor at 300 Hz? Neglect damping.

4.   A microphone is positioned 6 m from a noisy gearbox which acts as a point source and repeats a noise cycle exactly. A cancellation loudspeaker is positioned 2 m from the gearbox in line with the microphone and is programmed to give exact sound cancellation at the detection microphone.

Determine the distances from the microphone it is possible to move where sound level will be reduced by 6 dB or more from the original sound level at the microphone without correction. The wavelength of the sound is 1 m (i.e., 300 Hz) and it should be assumed that there are no reflections so that the inverse square law applies.

Chapter 13

1.   A gearbox runs for a million revolutions and is then stripped down for inspection. What action is advisable if you find
    a)   Limited pitting near the pitch line?
    b)   Pitting over the whole surface?
    c)   Wear of the whole gear face?
    d)   Fine cracks in the tooth root?
    e)   Fine pitting on the pitch line?
    f)   Tearing and smearing of the surface?

g) Deformation of the tooth flank?

h) Pitting at one end?

Chapter 14

1. A pinion of moment of inertia 1 kgm² meshes with a large and massive wheel which may be assumed not to vibrate. The base circle radius is 125 mm, the facewidth is 250 mm and the contact ratio is 1.5 with low helix angle. The main component of the T.E. is at once per tooth frequency and is sinusoidal with an amplitude of ±10 microns. The pinion is driven through a shaft which is very flexible torsionally but carries a steady torque of 500 Nm; bearing and shaft lateral deflections are negligible.

At what tooth frequency would you expect contact to be lost? What maximum amplitude of vibration would you expect to encounter as speed is reduced through the resonance in the system if the coefficient of restitution at tooth contact is 0.6?

Chapter 15

1. Determine the sizes of two possible 4 to 1 ratio final drives transmitting 10,000 kW at an output speed of 120 rpm. Assume that four planets are used for an epicyclic drive and that for a parallel shaft design the power is split onto two final drive pinions. Pinions and planets are "square", i.e., their length equals their pitch diameter and they may be loaded to 60 N/mm facewidth for each mm of module, i.e., 6 mm module gears may be loaded to 360 N/mm and other sizes pro-rata. Not less than 22 teeth should be used on any gear. Neglect helical effects and assume standard 20° teeth.

2. Compare the volumes and diameters of two epicyclic gearboxes, one of which uses three planets and the other uses five planets for a 4:1 reduction ratio, assuming that permissible loading per unit facewidth is proportional to tooth size and hence to diameter. Both boxes use "square" planets.

3. A compound planet spur reduction gear has a first stage sun wheel with 20 teeth meshing with three planets each of which has 52 teeth. The 1st stage planets are integral with the second stage planets which have 20 teeth and the planet carrier is the output. The second stage has a fixed annulus with 68 teeth. All teeth are standard 20° angle.

What is the overall reduction ratio and what is the permissible power at an output speed of 100 rpm if the second stage planets are 6 mm module, 75 mm facewidth and can be stressed to 280 N per mm facewidth? Could the overall ratio be increased by altering the first stage gears?

4. A split power system as sketched in Figure 15.6 uses an hydraulic drive which can transmit 40% of the input engine power but has an efficiency of 85%. Losses in the epicyclic gear may be neglected and the input speed and power are constant.

What are the highest and lowest speed of the output, expressed as percentages of the output speed when the annulus is locked?

5. A journey of 35 miles requires an average power of 15 kW for an hour. What weight of petrol, battery, flywheel, or spring stored energy would be needed for this requirement?

A mobile library weighs 5 tons including batteries which can give 250 amps at 96 V. The rolling resistance is 1500 N and air resistance may be neglected.

What is the maximum speed on level ground and on a gradient of 1 in 20? If the motor control system limits the propulsive force at the wheels to 5,000 N, how long will it take to accelerate to 4 m/s (about 10 mph) on level ground? Neglect losses.

ANSWERS TO PROBLEMS

Chapter 2

1. 2.031, 1.698, 1.500
2. 0.625 m, 14.08°
3. 15.09, 1.704, 9.62 mm
4. 1.796, 612 mm/s, 1.224 kW, yes
5. 0.773 m, 2.546 m
6. 1.602, 0.664 m, 23.88°
7. (a) 111 kW, (b) 119.7%, 69%
8. (a) 26.5% increase in diameter, (b) 33.3%

Chapter 3

1. 163.85 1.485, 748.4 Nm, 15.62°
2. 23.896°, 21.707°, 22.374°, 118.8 N/mm
3. 3 mm, 32.462°, 23.334°, 1.337, 60.031 mm, 61.935 mm, 65.685 mm, 436.5 Nm
4. 2.25 mm, 25.482°, 61.728 mm, 10.556 mm, 9.737 mm, 71.85 N/mm, 66.28 N/mm
5. 30.168°, 26 teeth, 43.985 mm, 57.500 mm, 1.9% increase
6. Pitch helix 16.754°, 5 mm module, 79.4% increase
7. (a) 0.728, (b) 1.367
8. 39.5 teeth, 19.75 teeth

Chapter 7

1. 37 microns, $1.57°$ of roll angle
2. 220 N/mm
3. 10.8 microns tilt and 7.3 microns crowning
4. 51.6 seconds of arc
5. 32.5 seconds of arc
6. 7.9 microns, 3.85 microns

Chapter 8

1. 15.6 N, 15.6 N, 15.6 N,
   6.8 N, 7.5 N, 17.7 N,
   506 N, 62.5 N, 9.1 N
2. 90 N, 202 N, Negligible, 10% and 6% decrease
3. (a) 159 N through wheel and 71.2 N, i.e., 7 dB reduction

Chapter 10

1. ω^2 is 7.8×10^6, 1.7×10^7, 2.6×10^7, 7.4×10^8
2. ω^2 is 2.1×10^7, 4.6×10^8 then 2.06×10^7, 2.6×10^8
3. ω^2 is 2.7×10^6, 4.1×10^6, 3.5×10^7, 4.5×10^8 for steel
   2.6×10^6, 4.1×10^6, 2.8×10^7, 5.0×10^7 for nylon
4. 0.138 micron
5. 0.093 micron, 83.3 N
6. 507 Hz, 1.4×10^9 N/m

Chapter 11

1. 528 N, 70.7 N. Leave pinion as it is and reduce wheel stiffness to 5.7×10^7 N/m, giving forces of 23 and 309 N
2. 5.03 Hz and 223.5 Hz (and multiples)
3. 10^8 N/m below subplate and 3.4×10^9 below gearbox
   Attenuation about 35, i.e. 31 dB
4. 1.23 m in front of microphone and 1.4 m either side

Chapter 13

1. (a) Nothing, (b) Overstressed, redesign, (c) Clean oil, (d) Stop immediately, redesign, (e) Earth electrostatic, (f) Harden surface, use EP oil, (g) Harder material, (h) Realign

Chapter 14

1. 398 Hz, 55.6 microns

Chapter 15

1. Wheel 1760 dia × 440 mm face
   Annulus 1056 dia × 352 mm face width
2. 66.7% volume and 18.6% diameter increase for 3 planets
3. 9.84, 140.8 kW. Yes slight increase by using 3.5 mm module gears but tooth
   correction would be needed to achieve center distances. 3 mm module gears
   would fit but would be overstressed
4. 156.7% and 70.1% (but losses only 6%)
5. 5.4 kg, 500 kg, 3.75 tons, 1800 tons. 16 m/s, 6.07 m/s, 5.7 s

INDEX

A
Addendum, 9
Averaging, 92

B
Backlash, 9
Base circles, 7
Bedding checks, 54
Bending correction, 71
Bevel gears, 31
Bouncing, 142

C
Cancellation, 128
Contact line lengths, 23
Contact ratio, 9
Contact stiffness, 61
Contact stresses, spur, 11
Correlation, 90
Couplings, gear, 36
Cracking, root, 137
Cracking, case, 139

D
Damage, 135
Damping, 109, 124
Damping absorbers, 125, 146
Discharge, 139

E
End relief, 63
Epicyclic, 149
Epicyclic, tabular method, 150

F
Factors, stress, 134
Filters, 87
Fourier analysis, 87

G
Gregory method curvature, 27

H
Helical reasons, 19
Helical gear relationships, 23
Hertzian stresses, 11, 26
Hypoid gears, 31

I
Idealization of system, 78, 103, 108
Internal gears, 15
Involute, choice of, 3
Isolation, 126

J
Jitter, 96
Jumps, 141

L
Large systems, 113
Load distribution estimation, 112
Load sharing, 148, 153

M
Maag grinding, 45
Modelling, matrixes, 103, 108
   scale, 116